T0181761

Studies in Computational Intelligence

Volume 745

Series editor

Janusz Kacprzyk, Polish Academy of Sciences, Warsaw, Poland
e-mail: kacprzyk@ibspan.waw.pl

About this Series

The series "Studies in Computational Intelligence" (SCI) publishes new developments and advances in the various areas of computational intelligence—quickly and with a high quality. The intent is to cover the theory, applications, and design methods of computational intelligence, as embedded in the fields of engineering, computer science, physics and life sciences, as well as the methodologies behind them. The series contains monographs, lecture notes and edited volumes in computational intelligence spanning the areas of neural networks, connectionist systems, genetic algorithms, evolutionary computation, artificial intelligence, cellular automata, self-organizing systems, soft computing, fuzzy systems, and hybrid intelligent systems. Of particular value to both the contributors and the readership are the short publication timeframe and the worldwide distribution, which enable both wide and rapid dissemination of research output.

More information about this series at http://www.springer.com/series/7092

Matthew Montebello

AI Injected e-Learning

The Future of Online Education

 Springer

Matthew Montebello
Department of Artificial Intelligence
University of Malta
Msida
Malta

ISSN 1860-949X   ISSN 1860-9503   (electronic)
Studies in Computational Intelligence
ISBN 978-3-319-88513-1   ISBN 978-3-319-67928-0   (eBook)
https://doi.org/10.1007/978-3-319-67928-0

Printed on acid-free paper

This Springer imprint is published by Springer Nature
The registered company is Springer International Publishing AG
The registered company address is: Gewerbestrasse 11, 6330 Cham, Switzerland

I dedicate this work to my parents
Joe *and* Grace
to whom I owe all and who still provide
care, support, and unconditional love.

Foreword

How do we plan for the future of education? Will it be vastly different from contemporary education, or will it continue in much the same way? What role will e-learning, computers and artificial intelligence play in that future? Will they be instrumental in transforming our students' experiences into more engaging, authentic learning, or will they perpetuate the same pedagogies and practices of old? These are not easy questions, because they are about the future, and as the old Norse saying goes: Predictions are difficult, especially if they are about the future. In this book, Matthew Montebello will show you that to look forward you first need to look back. The trajectory of e-learning is exciting, confused, patchy and multi-directional. It has all the twists and turns of a Game of Thrones episode, but without the bloodshed and nudity. This book examines the history of technology supported learning, and reveals some of the nuances and intricacies that make it what it is today widely adopted, yet poorly understood. Just like Game of Thrones, the plot has multiple dimensions, but ultimately, all of the storylines draw together and make sense as Matthew Montebello accompanies us on a journey to better understand what e-learning is really all about, and how it might pan out in the future.

One of the things I like about Matthew Montebello is the stance he takes on evaluating ideas and balancing arguments. In this book, he casts his critical eye over just about every aspect of technology supported learning, bravely thrusts his theoretical sword into numerous hornet nests and emerges with some thoughtful conclusions about what works and what does not. This is quite a balancing act when we consider the impasse that machines and humans have reached. Computers are logical and follow instructions precisely, and without deviation. There is no chance that they will ever break a rule or transgress the routines they are tasked to follow. At a fundamental level, this is the basis of the Artificial Intelligence systems (AI) Matthew features in this book. They are unthinking, unfeeling and blindly loyal to the codes that regulate their functions a little like the White Walkers. In contrast, humans are emotional and follow their intuition, and tend to bend or even break the rules if they see fit to do so. The interface between the two and the user experience that results, gives us pause to consider the nature of the human mind, the nature of machine intelligence, and the often uneasy relationship between the two.

The human aspects of education are rich, empathetic and emotional completely and starkly in contrast to the characteristics a computer exhibits. And herein lies the conundrum: To what aspects of pedagogy can machines be successfully applied, and what roles remain for human educators? What is the place for machines and humans in future learning and teaching scenarios? Indeed, we all have a view on AI and its potential roles in the future of education, lifelong learning and society. Whether it has reached its potential and transformed any aspect of education is for the reader to decide, but Matthew argues that to date, AI has a poor performance record, especially in technology mediated forms of learning support.

To date, computers have been deployed to manage all of the mundane and repetitive tasks that previously gouged huge chunks out of a teachers time. Computers can also make life easier for teachers if they are used as mindtools to offload much of the cognitive load, freeing up more creative thinking time. At a push, computers can even be designed and programmed to engage with students and to challenge their thinking by presenting and representing knowledge at a very high level. But can, or should computers (in the form of AI) replace any of the human functions teachers currently perform? If the answer is yes, what might be the implications for pedagogy, for students, for culture and ultimately for our wider society? And if AI was deployed in a grand, sweeping way to influence all aspects of pedagogy, would this actually result in any significant transformation of education? Has any previous application of technology created lasting and welcome changes to education? Such questions and many similar debates feature throughout the pages of Matthews book. Although not all the answers are provided, he certainly raises the profile of these important questions, and contextualises them in a manner that is both accessible and thought provoking.

So, mount that dragon, draw your sword and get ready for an interesting ride. There will be no White Walkers to battle with, but the machines are on their way— and sides are being taken.

Plymouth, UK   Prof. Steve Wheeler
August 2017

Preface

The idea behind this first book stemmed from a combination of ideas about e-learning and personalisation together with adopted concepts from a research initiative about crowdsourcing. E-learning has long been a domain which attracted my attention as I professionally chose to work within the areas of education at the outset of my career and eventually switched to computer science as personal computers made their public appearance in the late 80s. Applying information technology to education brought together both interests into one focussed domain that turned out to be a vibrant and flourishing domain that required further research, experimentation and development.

What started as an undergraduate assignment at the University of Malta in 2014 resulted in a number of prototypes that explored the notion of combining a number of complementary techniques, that have been successfully employed in a variety of domains, in an attempt to enhance the effectiveness of e-learning. The project took a major twist in 2015 when a decision was taken of blending three practices or approaches to potentially take e-learning to the next level by effectively adding value through the process of customisation. In the Summer of that same year, a fully fledged empirical study was held in collaboration with the national education authority in Malta to thoroughly test and collect numerous results from a mixture of data collection methods. The final outcome helped shed light on the techniques employed, the methodologies adopted and the philosophical reasoning behind it that have now been captured in this book.

Apart from the expected build-up to the proposed e-learning model whereby numerous subject matters from education and computer science are tackled, I wanted the reader to understand and appreciate how all these fit in and how e-learning has come a long way since its inception. Understanding how the parallel evolutions of technology and e-learning came about is a key to better position this work and comprehend the need for the next generation of online learning to materialise. The concepts, reasoning and rationale behind the model is, what is being presented with the knowledge that this has been tangibly deployed and tested in real life with higher education learners as part of their continuous professional development. The book is intended to lead e-learning researchers to further their efforts and work to improve e-learning effectiveness as they attempt to develop next-generation platforms in line with the needs of a modern twenty-first century connected society.

During the actual writing of the book, I had numerous occasions to look back at the empirical study and extract further insights into the model especially during discussions held with colleagues, peers and interested academics that I met at numerous conferences, where I presented my ideas and the rationale behind it all. The feedback I got was outstanding as the model presented a different perspective and fresh outlook on the evolution of how we perform online education. This positive response and constructive criticism helped me further refine the model and present a much more coherent and stable version of how I envisage the future of e-learning to be.

This experience helped me to look within and switch my role from an AI academic, researcher and educator to a raconteur and remote reporter looking for the most effective way to relay back the thinking process without integrating too much technical details that would otherwise over-complicate the pedagogical and philosophical reasoning behind the concepts presented, while maintaining my own personal epistemological beliefs to weigh in on my thoughts and writing.

The urge to pursue further detail in both the development aspect and the AI algorithms was a recurring challenge that, as a technical person I continuously was aware of but which avoided ensuring the proposed model stood up on its own merit away from the programming language or the specific machine learning technique employed when profiling the learner. The concepts and techniques can now be deployed independently from the specific technicalities that were employed during the empirical study.

Xaghra, Gozo   Matthew Montebello
July 2017

Acknowledgements

This work would have not been possible without the input given by my colleagues and students at the AI department within the University of Malta, as well as the invaluable contributions given by other academics at the School of Education within the University of Sheffield. Family and friends have played their instrumental roles in chasing, supporting and inquiring about the progress and completion of the book itself in these last 6 months.

Contents

About the Author

Matthew Montebello was born at Mtarfa in 1966 and attended school at St. Francis in Msida and St. Albert the Great college in Valletta before pursuing his education at the New Lyceum and the University of Malta. In 1990, he graduated as a Mathematics teacher and taught at a number of government schools while assisting the Ministry of Education to promote and introduce computer science in government schools. In 1995, he decided to further his studies in USA and UK with a masters and doctorate in Computer Science, and in 1999 he returned to Malta as a lecturer at the University of Malta where he was promoted to Senior Lecturer in 2004 and Associate Professor in 2013. His research in Artificial Intelligence and e-Learning is reflected in numerous publications locally and internationally in books, journals and conferences. He has also contributed to the individual and collective interests of the academic staff as a member, General Secretary and President of the academics union, UMASA. In 2013, he was instrumental in the successful signing of the University of Malta academics collective agreement with the Government of Malta. Professor Montebello also obtained a masters in Maritime Archaeology in 2008 and recently completed a doctorate in Education. He is heavily involved in voluntary work in Malta and Gozo, and sits on numerous boards locally at university level and abroad within the European Commission and international academic bodies. Professor Montebello is also an avid scuba diver with professional achievements as an instructor trainer within the technical and rebreather diving field, while contributing on a national scale to the training of Army and Civil Protection personnel, as well as a court expert in the area.

Acronyms

AI	Artificial Intelligence
AmI	Ambient Intelligence
AmILE	Ambient Intelligent Learning Environment
CGI	Common Gateway Interface
DTD	Document Type Definition
FTP	File Transfer Protocol
HTML	HyperText Mark-up Language
IoT	Internet of Things
LMS	Learning Management System
OER	Open Educational Resources
OSI	Open-Source Initiative
PAN	Personal Area Network
PLE	Personal Learning Environment
PLN	Personal Learning Network
PLP	Personal Learning Portfolio
RDF	Resource Description Framework
TCP/IP	Transmission Control Protocol/Internet Protocol
W3C	World-Wide Web Consortium
Web 2.0	Second generation Web technologies
WWW	World-Wide Web

List of Figures

Chapter 1
Introduction

Start by doing what's necessary;
then do what's possible;
and suddenly you are doing the impossible.

Francis of Assisi

The areas of Artificial Intelligence (AI) and e-Learning from the Computer Science and Education domains respectively are not usually associated together, not because they are not compatible or complementary but due to a number of other non-technical reasons. Considering that e-learning emerged from its distance learning predecessor, the concept of uniquely adapting the academic content to fit every individual student was not an option and providers did not even see the need to do so. Educational programmes and materials were designed and prepared in a standard and one size fits all way as they would have been for a face-to-face environment thereby the idea of employing any kind of intelligent computations was inconceivable and unnecessary. Apart from this fact there has always been the ideological block that learning, education and teaching can never be replaced by a machine and thereby the idea of applying any form of artificial treatment to a domain that should only be entrusted to trained professionals was discouraged, criticised and discounted.

Another factor that brought about such a line of thought was the fact that AI had been established well before e-learning was formulated and yet AI failed to achieve its potential as e-learning set off, blossomed and flourished leaving AI behind. This reflected badly on AI which did not rise to the occasion and certainly was not taken seriously enough to be entrusted with the complexities and capabilities that are usually taken for granted when performed by an educator. Since then AI has matured, applied extensively, and established itself within numerous domains that similar to humans can identify intricate and innovative solutions to issues that have never been examined before, as well as offer clarifications to complex and ambiguous situations which we thought machines could have never been capable of. It is time to entrust the complexities of learner profiling to established automated systems that can assist and not replace human expertise within an educational environment. The ability to employ such systems is crucial and deterministic in its success and its successful

© Springer International Publishing AG 2018
M. Montebello, *AI Injected e-Learning*, Studies in Computational Intelligence 745,
https://doi.org/10.1007/978-3-319-67928-0_1

adoption, but most of all in its trust by academic practitioners and administrators alike.

If history had to repeat itself and AI fails to deliver what computer scientists have been promising then the credibility of AI and the predicted benefits from its application will collapse and collapse even stronger while being discounted for good. The e-learning arena has also evolved significantly and has developed good practices together with optimal design guidelines, but still one can not say that it is a perfect science and a certainty, rendering the integration of additional variables like AI not as straight forward as one would like to. Numerous studies have highlighted the shortcomings of e-learning as we shall see in the following chapters, and in order to look ahead to future generations we need to ensure that we address such issues and weaknesses. Apart from the technological challenges that automating an educational environment involves, additional social, psychological and philosophical concerns need to be addressed and handled. Education is unlike technical and crisp scientific domains like Mathematics, Physics and Computer Science where AI has been successfully applied over and over again and yet such systems still encounter practical uncertainties and difficulties when employed in conjunction with people or within a social setting. Even developing e-learning systems for such domains still does not guarantee that the pedagogical processes employed will function seemlessly without any glitches or the certainty that it will be successful. Even traditional face-to-face instruction can have setbacks and occasional predicaments depending on the methodology adopted, the technique employed, resources and teaching aids utilised, as well as the educator and even the learners. So let alone automated e-learning systems that have been programmed to assist or even replicate the face-to-face equivalent.

Apart from the pedagogical aspect one also needs to keep into consideration content aspects as well as contextual implications related to the target audience. Educational practitioners together with learning theorists and academic researchers have long been discussing, arguing and deliberating about optimal ways on how to teach and deliver educational material over the e-learning medium, and yet they cannot categorically state or collectively agree that online education has reached a terminal perfection that requires no enhancements or can not be improved in any way. New solutions are required to address known dilemmas and what better way to productively adopt favourable and promising technologies that can alleviate and relieve some of the issues. Such technologies can potentially complement and assist educators who are required to provide the necessary human touch while ensuring that personal and individual learner characteristics are taken into consideration throughout the e-learning process. The quest turns into an imperative task of determining how much and in which ways to amalgamate and merge the two in a way to advantageously optimise the entire process. A trade-off between one and the other does not necessarily mean that once this is reached it has to imposed on all the learners, but a much more realistic and pragmatic way would be that the continuum is a variable and possibly customisable to the specific preferences and unique learning approach of each student. Learning styles differ from one students preference to another, just as a variety of instructional strategies suit different learners differently as what works with one learner does not necessarily work with another. However, at the end of each educational endeavour the educators, e-learning administrators and the educational institution will be responsible for the eventual outcome, while the supporting technology and the embedded intelligence will remain what they realistically are, pedagogic tools and teaching aids which simply reflect the academic decisions taken by the persons responsible.

From a different point of view away from the technical and philosophical reasoning that surround e-learning and how to enhance such a medium, there exist those sceptics that resist the technology and fail to trust what benefits can be extracted. Compromising learner privacy is one of the strongest arguments not to entrust confidential and private information with indifferent computerised systems that have no regard to the human learner at all. Such a pretext is unfounded if the function of the automated software is to enhance the educational experience and assist the learner to improve and ameliorate the academic progress. However other reasons like lack of control, poor self-confidence, and fear of being over-powered by an unknown entity, can all be brought into play as a valid excuse to reduce learners dependence on technology as well as ensuring that educators will not be replaced. Technology and artificial intelligence are not meant to diminish or replace any learners or educators skills or capabilities. Such academic aids and utilities are not intended to transform education into a commodity or an automated programmed process, but merely to support, complement, and intensify the benefits of education. How best to optimise the use of such tools, technologies, and techniques to enrich, enhance and ameliorate their service and contribution to e-learning in general? What role should they play within the bigger e-learning picture? Should the learner or the educator have the liberty to be able to tweak them? Should the role if these applications be that of decision support systems or major or even sole contributors? How will they evolve as the technology itself evolves and still ensure they provide the intended academic benefits?

These are the questions this book sets out to answer as each chapter incrementally and sequentially builds on the previous towards a proposed future generation of e-learning. It is essential to comprehend and appreciate where e-learning is coming from and the evolutionary course it went through while shadowing the amazing technological progression. Chapter 2 gives an in-depth analysis of this journey as it offers numerous perspective evaluations of different influential e-learning factors and players. A full array of latest and relevant technologies are placed into context in the next three chapters that can be read in isolation but that are instrumental in the way e-learning will mature, renovate itself, and foster new and innovative interest. Topics include socially relevant issues like MOOCs, crowdsourcing and social networks, as well as AI research areas of user profiling and personalisation, together with e-learning literature related to personal learning networks, protfolios and environments. Chapter 6 brings the previous chapters together as it presents a model of how the next generation e-learning could potentially be like, as a combination of technologies grounded within respective learning theories come gracefully and compatibly together complementing each other while addressing numerous e-learning issues presented earlier. Finally the book closes with a predictive look-ahead of how such a model can materialise as the numerous interposing factors and contributing players are critically evaluated and intensely analysed as they possibly could change and characterise the future of e-learning.

Chapter 2
e-Learning so Far

If I have seen further,
it is by standing on
the shoulders of giants.

Sir Isaac Newton

Abstract e-learning has come a long way and it is only thanks to previous versions and numerous evolutions of e-learning that we can propose new routes and design intelligent systems for future generations. This also enables us to appreciate and value the meaning of moving forward as we fully understand and acknowledge from where we are coming. A plethora of research studies have reported conflicting results over the years as some praise and applaud this medium, while others disapprove and critise e-learning in every possible way. The fact that e-learning itself is not regulated by a specific academic body and that best practices are subjective, divergent and too generic, renders the whole playing game fuzzy, confusing and incredibly frustrating to the learners. However, pedagogical trends and technological forces have shaped the history of e-learning and will continue to do so. How have these rubbed off onto each other? And how have they influenced the following generation of e-learning? What are the factors that will impinge on the future of online education? In this chapter a deeper examination and appreciation of these changes and developments over the years is presented in an effort to understand the inevitable evolution that occurred and how this affected and influenced the whole environment surrounding e-learning. These include the social implications, the pedagogical repercussions and the technological impacts that gave rise to different e-learning generations.

2.1 e-Learning Generations

It is important to distinguish e-learning from its predecessor distance learning that can be traced back to the 1700s. Even though not electronic it still exhibited the characteristics of non face-to-face education that did not involve humans directly. The actual electronic characteristic is what distinguishes e-learning from correspondence courses, education through post, telephone, radio and television broadcasts up until the 1970s. The fact that a different medium was being employed instead of the traditional and classical student-teacher interaction marks a departure point whereby, irrespective of the actual medium, education was being packaged for other students who for some reason or other are not in the vicinity of the educator. This demarcation in itself that simply alters a single variable, the physical location of student and teacher, effected the entire educational process as theorists and practitioners have argued and debated over the years. The fact that the educational process is a complex one and not universally defined or specified justifies the intricacies created as soon as a single variable was modified. To keep in mind that even without the alteration of this single variable the student-teacher relationship and the learning process were not always optimal and even though numerous learning theories postulated the soundest methodologies to employ, the final product left much to be desired. To such ends distance learning and eventually e-learning inherited such challenges in addition to an already herculean mission to simulate and match the human counterpart. If this was not enough a supplementary technological overhead intensifies the issues and challenges that need to be overcome. The electronic factor mentioned earlier that differentiates distance learning from e-learning is nothing more than the technological component that enabled the shift from distance learning to the first generation of e-learning. Technological evolutions that over the years have enabled the web to develop further have surely altered the medium employed and as a consequence provoked a respective development and expected progression in the e-learning camp.

2.1.1 Taking Distance Learning into the Electronic Age

The network and the existence of an electronic connection was surely the beginning of a multitude of concepts that allowed the transportation of information between two nodes. DARPA, the Defence Advanced Research Projects Agency in the United States of America spearheaded this initiative and by 1969 Arpanet was the first network that enabled the sharing of educational and research material. The medium over which learning material was being passed on and shared or traded had just been upgraded from mail to electronic networks. This same electronic network was the one to evolve and embrace the implementation of the standardised TCP/IP, the Transmission Control Protocol (TCP) and the Internet Protocol (IP), in 1974. These international protocols provided the ideal and safe connection to communicate and distribute educational material from an educator or an institution providing training to any learners who were connected over the same network. The communication part was mainly done through basic electronic mail or e-mail, while transferring of files was regulated using another important standard or protocol called FTP, File Transfer Protocol. This period demarcates the beginning of e-learning whereby electronic means are engaged to teach and employed as a medium to pass on educational material to learners. Simple as it may be this primitive and rudimentary e-learning conveniently served its purpose and remains at the centre of all future generations of e-learning.

2.1.2 e-Learning Through the HTML Era

As the web itself evolved from its first generation of solely network connectivity allowing academics and learners to communicate through e-mail and share resources through FTP, to the extensive use of HTML (Hyper-Text Markup Language) to display webpages. This richer interface allowed web users to browse through the different pages that mainly depicted hyperlinks to other documents and images through the use of web browsers. This was an era where few search engines and humanly compiled directory services attempted to take stock of all the webpages while connected users from any connected node started pouring masses and masses of heterogeneous marked-up pages interlinked and unstructured. Web crawlers took advantage of such hyperlinks to scan documents in an attempt to parse and index every single document and webpage that web browsers had access to. Webpage authors, especially those whose intentions were to feature high on the search engines top ten results, employed tactics and techniques to benefit from the unsophisticated web crawlers software as explicit meta-language details were mainly employed for indexing purposes.

Another important aspect of this era was the integration of both presentation and content together without any proper distinction of one from the other within the same webpage. The only way web crawlers distinguished between content and anything else was when encountering a <while parsing the webpage. This mainly meant that an HTML presentation tag was about to start and thereby not of importance to a crawler which resumed its filtering once the closing> closed the HTML tag. However even though this worked for some time web page authors quickly realised how to deceive the crawling software. The consequences of such a configuration with the content and presentation merged together were quite significant and which would characterise this particular web generation. Apart from the fact that HTML is a loosely-typed language, the distinction of where information or data starts and ends, and the details of the mark-up that pertained to the actual webpage visuals and presentation through the browser were blurred and intertwined. Furthermore, different proprietary web browsers had conflicting configuration procedures creating confusion amongst web authors and dissatisfaction and perplexity amongst web users. This was initially resolved through the arduous authoring different HTML documents for the major web browsers, and eventually thanks to the use of the Javascript scripting language it was possible to distinguish between the hosting web browser and correctly display the contents of the HTML page to the respective web browser. However, this did not resolve the issue of having the same HTML page display on different output devices especially as technology evolved and webpages were able to be displayed over a variety of devices. The fact that the academic content and the presentation of the educational environment that was to be displayed on the learners browser was merged together created issues, repetitive work and major frustration to e-learning authoring bodies.

Another aspect of this era was the introduction of some dynamic feedback and interactive effect once a user or a learner submits personal or logging in information. This was possible through the use of the Common Gateway Interface (CGI) protocol that started and matured in the mid 1990s and which necessarily allowed web pages uploaded on a server to ask users to submit information via a form which on reaching the server it is able to execute a computer program whose output was actually a fully-fledged web page with customised details as part of this same output. Such a dynamic novelty was exactly how the search engines returned their results but this time, especially login sites, where able to personalise the content and presentation to the unique user details. This was possible as the server-side computer programme, through the CGI protocol, accessed the data saved on the server side, potentially a database, and dynamically included the user specific data, results, requests and academic record for example as part of the HTML web page output. As soon as the server dynamically generated the web page and sent it back to the requesting browser it was displayed as a standard HTML page. Even though CGI scripts were relatively easy to develop and deploy as they were language independent, they were quite slow and used a considerable amount of processing power as each request to the server a new and separate HTTP request on the server had to be done, execute the program, access the database, and output the HTML response. During the same period another methodology, Cookies, were employed to customise the content of the web pages that were visually rendered on the clients web browser. Web cookies, on the contrary to the CGI scripts, are client-side based as textual information about the user that has been issued by the web site being browsed is saved on the same users computer. Each piece of information is relevant to that specific web page and serves as a technique to persistently recall personal information respective to that web page, like logging details, progress in a course, file references, etc. Web cookies are easy to implement and do not occupy a lot of space on the clients computer, however, they have numerous restrictions and totally in control of the user rather than the web site developer. Cookies are domain specific with a limited size and lack the flexibility and capacity to store large amounts of user information, nevertheless ideal to store basic user information that would otherwise require the user to memorise or record elsewhere.

In both cases, the use of CGI scripts and/or web cookies managed to inject a taste of customisation as they contributed snippets of personal information about the user, but at the end of either one of the processes an HTML web page was nonetheless outputted and rendered on the web browser? presentation and content still merged together.

2.1.3   Semantic Web Takes over

The dynamicity generated through the employment of the CGI protocol and web cookies was not enough to inject enough personalisation to take e-learning from its static state to a more attractive and effective personalised delivery. Additionally the fact that content and presentation were still fused together in an HTML document that rendered delivery and adaptation of the content laborious and problematic. If that was not enough, all this made searching, indexing and retrieving the exact material and content that a user is looking for or that an educational environment needed to make use of was unclear and inaccurate due to the unstructured, inexpressive and lacking meaning as well as context. The solution that surfaced towards the end of the 1990s was the use of another mark-up language, XML, or the eXtensible Mark-up Language. Even though XML is a mark-up like HTML, it is a lower-level tightly-types language which has a stricter set of rules to encode documents that are still readable by people but most importantly can be parsed by a computer or machine-readable. The beauty of XML includes the distinction between the presentation and the content and this means that once the educational content is authored and generated it can be rendered and presented numerous times on different platforms and devices by simply employing the respective presentation medium to render the same content. Another benefit of XML is that it is extensible with the potential to generate appropriate and relevant tags that are precisely fitting to the content that they describe. An XML document with precisely tagged content within is required to be well-formed and valid before it is published. To such extents XML documents have to strictly satisfy all the syntactical rules specified by the World-Wide Web Consortium (W3C) that internationally regulates the use of XML and all other web standards. Additionally the XML document is required to follow the structure or scheme as defined within an accompanying document called the Document Type Definition (DFD) in order for it to be validated and published. The DFD specifies what element and attributes can be used together with any grammatical rules that need to be followed. This is what makes XML documents so expressive and useful as they enable parsers, not only humans, to make sense out of them and ensure that they relate to exactly what they are meant to be associated with. XML enabled and assisted the setting up of the Semantic Web concept, the web with a meaning. To such extent any educational material be it courses, resources, content, and any other academic matter can be meaningful used and reused by online system without the direct intervention of people. This is possible through the effective use of the DFD files that contain specific elements with their respective attributes, which in this case will be associated to an educational setting. The way this is possible is through the use of Ontologies that represent the position and use of the same elements referred to in the DFD to other academic terms and elements. A comprehensive education ontology explicitly species every conceivable academic term and describes how all these terms are related to each other in a tree-like way to simplify the meaningful and accurate use of the same academic terminology in a precise and coherent way.

In reality the Semantic web technologies as described above did not reach their expected potential as e-learning systems were not developed enough to employ software, like say pedagogical agents, to exploit the advantages of a machine-readable web. More on pedagogical agents and how the Semantic Web can still directly assist, facilitate and contribute to the evolution of e-learning will be discussed in Chap. 5. However the semantic web still added value to the web as everybody knew it as a web of meaning gave rise to the second generation web or Web 2.0 that carried with it related technologies and functionalities.

2.1.4   Empowering the Learner Through Social Networks

Much of the technologies related to Web 2.0 are socially enabled applications that empower and endow every web user by transforming each and every one of them from passive receivers and online content consumers to active and dynamic contributors. Learners have been transformed into authors and providers as they interact, socialise and share their work, experiences, and knowledge with other learners. The role of social networks within e-learning systems did not replace any conceptual or pedagogical notion from previous e-learning systems mentioned earlier, but took advantage of our human and social instincts to communicate with each other as learners and fellow students. The possibility of creating exclusive and focussed academic gatherings as well as public and unrestricted opportunities for others to contribute has never been possible within an e-learning environment. Students had no idea of other fellow course participants let alone sharing and contributing of ideas, offering feedback to each others work, and collaborating together to achieve a common goal. Such empowerment fosters bonding amongst learners that eventually supplements the learning process with additional interaction, knowledge exposure and a reassuring sense of achievement and accomplishment. Additionally through the use of ubiquitous hand-held devices online students can integrate the learning process seamlessly and effortlessly as part of their on-going daily activities by posting short messages, snippets or images of content related to the academic domain at hand, as well as bolster their rapport with their peers and trainers. Some networks enable learners and educators to reveal their professional profile and career achievements thereby promoting a hierarchy of knowledge and experience. This could add value to the e-learning process but tends to stir together the personal aspects of people with their role and participation in a course. Some other social networks, on the contrary, promote the anonymity of the participants by distinguishing between the learners and instructors personal social lives and their academic activities.

2.2 Technological Evolution Analysis

The evolution of the web and technology in general these last decades is amazing, breath taking and indisputably on-going. The parallel evolution of e-learning as it piggy-backed the technological advancements and the flourishing web has already been highlighted earlier in the chapter and acknowledged by numerous researchers amongst which are Hussain [1] and Miranda et al. [2] particularly underline the exact duality as the bi-evolutions are enumerated and categorised. The exact generations of both technologies, summarised in Fig. 2.1 are not only incrementally described and crisply interpreted in their own way, but also categorically distinguished from each other as if they were paradigm shifts that specifically happened at an exact point in time. In reality the evolution of technology evolved over long periods of time and eventually influenced the entire world around them including the way e-learning was performed. This does not mean that the technological advancement had a direct and immediate effect on e-learning systems but simply that the technological environment within which the e-learning platforms were implanted in changed and transformed into something else over time. The e-learning environment did not necessarily change due to the technological transformation and not did it adjust simply because a new technology was now around, but mainly due to functional and operational enhancements that rendered e-learning either more effective or easier to conduct. Intrinsically the quest has always been to simulate a human teacher and diminish as much as possible the rift between computer-based education and face-to-face delivery. The use of technology was not only intended as a teaching aid within an academic environment, but mainly to optimise the electronic medium being employed, and that

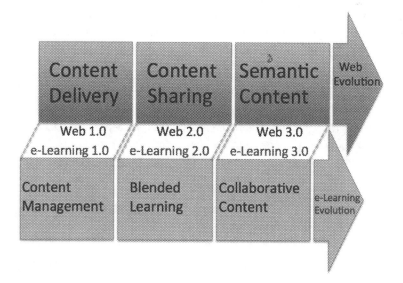

Fig. 2.1 Web and e-learning generations adapted from [1, 2]

is replacing the classical way educator and learners communicate. The point being made here is that technology led e-learning and the evolution of hardware, software and communication predisposed e-learning to conveniently adopt them in an attempt to improve and progress. Very few instances in the evolution of technology has e-learning imposed on the course of its development and perhaps it is proper to give tribute and praise the role of the University of Illinois (UoI) with its Illiac supercomputers and the different versions of the PLATO computer-based education system 1960s. Academics at UoI led by Donald Bitzer in the early 1960s designed the initial Illiac versions to specifically accommodate the requirements set by PLATO and thereby set a precedence of technology adapting to e-learning rather than the other way round. Ironically since then much of the e-learning evolutions simply happened as an adaptation to the continuous and relentless development in technology. Major shifts in technology that have definitely left their impact on e-learning have been highlighted at the beginning of this chapter. These technological evolutions invariably created a shift within the functionality of e-learning systems as a result of which it evolved they adapted, evolved and morphed over time. The network radically set the scene for e-learning and a solid foundation upon which future technologies could incrementally build layer upon layer. From HTML technology that allowed connected learners not only to communicate but access information, to XML that enabled meaningful interactivity thereby enhancing the educational environment, and eventually to Web 2.0 technologies that empowered e-learning students transforming them from simple recipients into creative and social contributors.

2.3   Unfolding Social Implications

The same technological conception of a network that launched the web as we know it today and triggered off e-learning per se, can be considered the inception of a society as it connected learners in one way or another who could communicate in rudimental ways. Even though e-learning was merely accessing educational information and downloading academic content, learners would have initially been enrolled as part of the educational institution student body and thereby part of a social group. Initially such a social group had no real communal interactions between the learners who barely knew about each other however one was admitted, enrolled and felt part of the educational institution. Eventually as the technology progressed it became possible to enable a higher level of interaction with fellow students who could relate to each other and communicate in some way amongst themselves apart from their educator. With the introduction of social networks and the adoption and integration of such media within e-learning environments the learners acquired a new dimension of communication. The social implications that unfolded as the technology progressed and evolved empowered learners but most of all energised e-learning to a levels never attained before. Students could now emerge from their isolated learning environments to collaborate together while interacting through a variety of media. Online social networks took the world by storm at the beginning of the 2000s and it was only natural that e-learning takes advantage leaving an incredible impact as it did on all matters related to society. Around 2005 education theorists like Siemens [3] and Downes [4] felt the need to propose an educational theory related to this contemporary pedagogical phenomenon which was coined as the Learning Theory of connectivism. The basis of this theory is founded upon the network technology and the capabilities enabled by networked computers. Due to its dependency on the electronic technology it has also been termed as a learning theory for the digital age as it relies on the simple and rapid collation of information and knowledge dispersed around the web and that is derived from knowledge bases and fellow online learners, experts and knowledge-providers. The importance of setting up and maintaining a healthy personal learning network is imperative for every learner who is now empowered to take control and manage the sequence of events and processes that determine ones own education.

As the technology evolved and e-learning matured the social state of affairs unfolded in a way that benefitted the learner but which indirectly impinged on the on-going educational process and its respective pedagogical practices.

2.4 Pedagogical Repurcussions

The state of e-learning pedagogies followed their face-to-face counterparts as e-learning was not even considered as a distinct discipline but simply an extension of traditional learning methodologies. Initially e-learning designers were inexistent and eventually as the domain matured and developed into an entity of its own designers attempted to optimise on previous versions of e-learning systems.

The classical behavioural and constructivist learning theories influenced and left their mark as technologically-inclined developers rather than educationally-minded theorists assisted in the first proper designs of e-learning systems. This can be evidenced in the typical e-learning platforms that simply transformed and regurgitated traditional learning courses into their electronic counterpart. Learners had access to their academic notes that they could download while accessing reading lists and following a sequential set of educational materials. Pedagogically such a scenario is no different from a face-to-face learning environment with the only difference of having access to the knowledge at any time, at any place, and as many times as they like. The pervasiveness that the technology enabled has no effect whatsoever on the pedagogical properties of e-learning. Similarly the availability of multimodal resources that the technology supported did not change the way an online course was being designed and delivered.

All this brings into perspective the fact that even though different technological evolutions occurred over the years together with sociological implications as interpreted in the previous sections, the e-learning pedagogies did not evolve or accommodate the transformations of the medium employed. Numerous researchers [5, 6] consider technology to be pedagogically neutral especially when the design is entirely focussed on the organisational structure of content rather than on the specific educational activities. The imbalance occurred when content was not only provided by the educator on the server-side but was now also dynamically provided from the learners client-side within the same e-learning environment. The educational impetus that Web 2.0 technologies brought about was tremendous with pedagogical ramifications that markedly challenged the traditional e-learning pedagogies to an extent that, as mentioned earlier, brought about the proposition of the constructivism learning theory to sustain such e-pedagogies.

This did not change the fact that a mixture of learning theories that sustain e-learning models and pedagogies that complement each other. A creative and pragmatic combination of learning theories that include behaviourism, constructivism, cognitivism and constructivism that support the best pedagogical practices that have been specifically prepared and distinctly intended to be employed within an e-learning environment.

2.5 Conclusion

In this chapter a deep analysis of how e-learning evolved over the years as a result of a parallel technological evolution has shown that the design and development of e-learning systems was unfortunately an afterthought. This will inevitably impinge on the overall success of these e-learning systems that had their share of issues and student concerns as will be pointed out in the following chapters. This book is an effort to address these e-learning issues and concerns by reversing the dysfunctional modus operandi of designing an e-learning system on the basis of the available technologies, and on the contrary strongly support and advocate the notion of adjusting and conceiving adequate technologies to optimise e-learning systems and their overall effectiveness.

References

1. Hussain, F.: E-Learning 3.0 = E-Learning 2.0 + Web 3.0?. In: IADIS International Conference on Cognition and Exploratory Learning in Digital Age (CELDA 2012), pp. 11–18. IADIS, Madrid (2012)
2. Miranda, P., Isaias, P., Costa, C.J.: E-learning and web generations: towards web 3.0 and E-learning 3.0. In: 2014 4th International Conference on Education, Research and Innovation, pp. 92–103. IACSIT, Singapore (2014)
3. Siemens, G.: Connectivism: A Learning Theory for The Digital Age. Elearnspace. http://www.elearnspace.org/Articles/connectivism.htm (2004). Accessed 22 Jun 2014
4. Downes, S.: An Introduction to Connective Knowledge. http://www.downes.ca/post/33034 (2005). Cited 25 Feb 2012
5. Cope, B., Kalantzis, M.: e-Learning Ecologies: Principles for New Learning and Assessment. Routledge, London (2017)
6. Hoel, T.: Standardizing e-learning: is the learning technology of tomorrow built on the learning theories of yesterday?. Thesis, IT University of Gteborg, Gteborg (2002)

Chapter 3
MOOCs, Crowdsourcing and Social Networks

Learn from the masses,
and then
teach them.

Mao Zedong

Abstract Social media took the world by storm and transformed the society and its multiple dimensions in more than one way. The extent of the shock waves that this phenomenon inevitably influenced the way people interacted with the web and with each other, as well as with all web applications and services provided online. E-learning evolved as it embraced the new Web 2.0 technologies in an attempt to enhance the delivery but at the same time to take full advantage as in the past of the latest cutting-edge technologies that were available. It has been argued in the previous chapter that this technological shift was no standard evolution but a major unconventional and progressive e-learning revolution that literally turned the tables around. In this chapter the full impact of this considerable technological contribution to the pedagogical and functional dynamics of e-learning will be brought into perspective as innovative techniques transpired from the evolution of Web 2.0 technologies that slowly but surely got integrated within online learning systems. These include MOOCs or Massive Online Open Courses, Crowdsourcing techniques, and Social Networks. The beauty about these technologies that resulted out of the latest technological evolution addressed particular e-learning concerns as e-learning had been emanating from the integration of a variety of incongruous emerging technologies that at the time assisted in improving the services provided by such systems. In the following sections the main e-learning issues will be discussed together with how emerging technologies can start addressing them.

© Springer International Publishing AG 2018
M. Montebello, *AI Injected e-Learning*, Studies in Computational Intelligence 745,
https://doi.org/10.1007/978-3-319-67928-0_3

15

3.1   e-Learning Effectiveness, Issues and Concerns

The improvised and unstructured way e-learning has evolved over the years ever since the network was set up resulted in an irregular yet understandable entity that was meant to perform a precise task but whose effectiveness was as successful as the accuracy it was measured by. The technological driving force that pushed and drove e-learning to develop into a functional and productive online commercial enterprise came with its own qualms that accumulated issues related to the educational processes employed as well as pedagogical concerns that needed to be addressed.

Numerous studies confirmed that online education did not achieve its potential and that there was no significant difference between the overall benefits of e-learning and the classical face-to-face interaction. However the need to clearly specify how to measure e-learning effectiveness has always been critical to determine whether specific methodologies, techniques and learning theories are more effective than others. Bloom [1] postulates through his 2-Sigma problem that individual education enhanced the learning effectiveness by 2 standard deviations. This obviously drives e-learning towards adopting such an individualistic approach in an attempt to enhance its effectiveness.

A variety of techniques have been employed over the years to measure and evaluate e-learning effectiveness as a number of research studies [2–4] reported a somewhat inconclusive and inconsistent findings. Piccoli et al., [5] similarly reported indefinite results when attempting to assess VLE effectiveness in relation to an educational programme of elementary ICT skills. Measuring specific e-learning characteristics to determine its effectiveness was a technique that a number of studies [6, 7] adopted as they consistently contrasted teaching over the electronic environment to the traditional face-to-face methodology. However the outcomes from these studies similar to the previous ones gave no definite indications as a mixture of outcomes reflected a lack of universal conformity of what an ideal e-learning environment should look like or whether it was more effective or not. In an attempt to harmonise such research in 2011 Academic Partnerships published a white paper [8] to catgorise the different e-learning research trends. The authors identify four distinct research areas whereby different aspects of e-learning are investigated, namely, academic outcomes of online education, growth of online learning, cost as a direct impact of its effectiveness, and impact on instructional design and delivery. The fourth of these e-learning effectiveness research domains deals directly with the topic of this book as the architectural setup and the pedagogical underlining determine the specific model being adopted when developing the e-learning platform. In a similar manner Chan et al., [9] employed a four-factor study to assess and evaluate e-learning effectiveness. They argued that e-learning is a complex entity involving numerous interrelated issues that need to be taken into consideration if one is to assess its effectiveness. To this effect they proposed the framework shown in Fig. 3.1 to assess and evaluate e-learning effectiveness and online education by taking into consideration the evaluation methods employed, the results that were achieved, as well as the course and its content and organization itself.

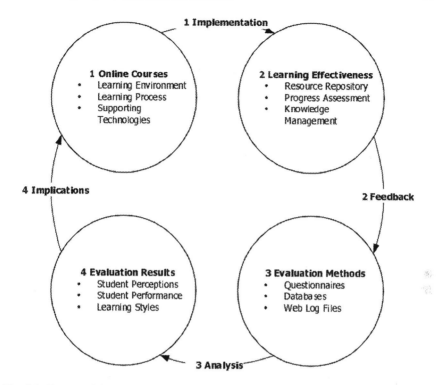

Fig. 3.1 Framework for evaluation of learning effectiveness in online courses [9]

The trend in mixed results reported when comparing e-learning effectiveness in contrast to face-to-face is also documented in other numerous research studies [10–13] however the potential of e-learning and the positive manner how these same studies come across have shown that e-learning is here to stay and that the availability aspect of such courses has helped in augmenting the rate of completion within the higher education domain. This reinforces the need of further investigation into how best to deploy e-learning platforms and research to assess the different methodologies and models that are developed and tested. The outcome of these studies also highlights the need of a proper and formally defined way of how to assess the effectiveness of e-learning courses in an attempt to identify and isolate those critical factors that require close and strict scrutiny. Neuhauser [14] designed a research study to reduce the number of variables and single out those e-learning factors that needed to be measured. Numerous best practices documented in this research included the duplication of two sections of the same academic content to ensure that all that varied were the intended measurable factors like learning preferences, study styles, methodology effectiveness, and overall environment valuation. The problem with similar studies is that they usually involve educators who already have a positive disposition towards technology and thereby tend to rub off onto their students creating an impartial bias within the study itself. On the other hand those learners who opted for e-learning independent from any influence tend to struggle in their attempt to adapt to the innovative learning processes and the responsibility of controlling their own progress and eventual academic success. Apart from the psychological challenge to adapt to a non-human led academic programme some learners encounter difficulties coming to terms with the technology itself, figuring out what needs to be done, ensuring they are not missing out on any academic content or task and figuring out what needs to be done to ensure they complete the required educational programme. How intuitive are e-learning systems? Have they always been designed to cater for learners hailing from the traditional classroom who are consistently hand-held and instructed what to do? Are the learners mature enough to follow the online programmes set without getting distracted, discouraged or even lost?

The proposed e-learning model presented in Chap. 6 addresses three of the most common e-learning challenges that have been at the focus of numerous researchers [15–17] as they individually attempted to address e-learning effectivenss by overcoming such concerns. Whereas some learners are disciplined and determined enough to follow an online course, others, similar to a classroom scenario, fall back and eventually fail the classor even drop out if they do not manage to administer their own time, activities and attitude towards their online education. Motivation plays a very important role and numerous e-learning critics have blamed it on the medium and its cold insensitive interface that fails to connect and build such a rapport that a human educator is able to establish with learners in class. Some students require continuous encouragement and motivationto maintain a correct attitude while following an e-learning course, while others are self-motivated and tend to perform well irrespective of the academic environment. Learner motivation could be affected by a number of issues but the lack of enthusiasm usually results from either learners who lack determination, or simply are not interested in the subject matter. Attempting to engage learners with the educational content by rendering it relevant to them and relate it as closely as possible to their own interests has been investigated by Tang and McCalla [18] where they highlight the importance of learner feedback in order to offer in return course materials that motivate further individual students based on their personal profile. Motivation is an important issue in every learning situation but in regards to e-learning the need for learners to be self-determined is even greater. In this respect the self-determination learning theory has occasionally been coupled with the corresponding learner profiling approach to address this particular issue.

Similarly some learners experience a sense of isolation [19] during an online course even though they interact electronically with others but still have a preference for the direct face-to-face interaction. Isolation refers to the learners lonely experience during an e-learning course without any contact whatsoever with other learners or educators. Bousaaid et al., [20] investigate this phenomenon and conclude that the simple act of participating within a network of like-interested persons within a social network assists e-learners and renders the entire process more effective. They argue that latest Web 2.0 technologies actually promote even more communal practices whereby learners are able to collaborate, share and communicate freely with others. Similarly Davies and Merchant [21] highlight the ability of Web 2.0 to enrich and transform the educational experience. Web 2.0 is considered to be a phase, or the second generation, of web technologies that promote user-generated content coupled with mechanisms that enable and enhance user interaction (p. 4). The authors identify four distinguishing characteristics to illustrate how web users can exploit Web 2.0, namely, through being present, the ability to modify and generate content, and finally by partaking in social activities. These features, apart from reducing the isolation problem, go further and promote the individuality of the users while establishing a personalisation element. The learning theory adopted to address this issue was Connectivism which, together with the approach and associated implementation, are addressed in the following chapters.

Finally, the third e-learning issue being addressed is that of e-learning being impersonal. Critics blame the impersonality of the technology for such concerns and the proposed model employs AI enabled learner profiling techniques to dynamically address such a concern together with the adaptive nature of Web 2.0 features. The adaptive learning theory is associated to this particular issue in tandem with per-sonalisation techniques as part of the solution presented in this book to address and enhance e-learning effectiveness. Will the technology live up to its expectations to address all these issues? Will the educational requirements lead the way to technol-ogy development? How will future e-learning systems manage to overcome such challenges?

3.2 MOOCs

Massive Open Online Courses have been established for quite some time since their initial appearance in 2008 and reached a peak four years later. The fact that these online courses are open and easily accessible made them very popular and gave the opportunity for anyone to enrol and follow. The concept behind MOOCs enabled free education for the masses bringing into perspective the desire of a lot of online users to learn more and which were restrained for some reason or another but mainly due to lack of accessibility, excessive fees or merely inconvenience. On the other hand MOOCs uncovered numerous issues that gave useful insights into e-learning in general. Apart from their poor retention rates and grading concerns it is practically impossible for an educator to provide assessment or even personalised feedback to each and every learner, and therefore the Web 2.0 technology provided the possibility for students to provide the required interaction as well as criticism, advice, opinions and useful tips. These technologies have also been successfully employed in standard e-learning courses that differ from MOOCs in a number of ways. However through MOOCs they have been further repurposed and fruitfully employed. Amongst others are tools like micro-blogs and wikis that enabled creativity, collaboration and coop-eration. Similarly use of social media, synchronous meetings and other Web 2.0 tools like polls, concerted authoring, and live surveys allowed learners, educators and any other online users to truly interact.

Even though MOOCs have been portrayed as the next silver bullet in e-learning, the concept and spirit in which they were conceived is what is important here. Siemens and Downes [22] launched their MOOC to research networked education in an unlinear crowdsourced way. Coursera [23], a popular MOOC platform, was set up by its originators Daphne Koller and Andrew Ng to grant the public access to the top academic materials and professors which were otherwise inaccessible. Similarly when the MOOC platform Udacity [24] was launched by its founder Sebastian Thrun the academic reason to do so was to reach out to his numerous students offering a personal experience in contrast to what they experienced within the over-flowing massive university lecture theatres. The essence of this takes us back to the issue of the academic needs imposing on the available technology rather then setting up a specific e-learning methodology simply because the technology is available or because the industry pressures and financial needs dictate so. The techniques, models and methodologies adopted by MOOCs have much to offer to the proposed model presenting opportunities for e-learning to extract the positive aspects of MOOCs while personalizing online learning for mass consumption.

3.3   Crowdsourcing

The concept behind crowdsourcing originated from industry as open source software and collective contributions from the masses gather momentum and constitute a force to be reckoned with. This technique enables the possibility and potential of bringing together diverse skills, competencies and expertise from all over the different online networks. Crowdsourcing is a successful tried and tested concept that the software domain [25] can vouch for through the Open-Source Initiative (OSI) whereby independent and freelance developers are more than pleased to share their expertise and contribute to a common final goal. A perfect example of such a phenomenon is the open encyclopaedia Wikipedia that accepts and values all kinds of contributions from online users who are able to append new information but also edit and author additional knowledge. The computer operating system Linux is also a great example of how crowdsourcing can enable multiple hundred developers to create an effective and efficient operating system. Other examples include the notorious Mechanical Turk [26] that facilitated the delegation of small tasks to complete a commissioned compound online piece of work.

The social aspect of aggregating online users to collectively combine their knowledge and expertise towards a common goal has been successfully and productively employed in other domains that include User interfaces [27], Soylent word-processor [28], Cultural applications [29], Commerce [30], Astronomy [31], News [32], Politics [33], and Employment applications like SuggestBot [34]. The common factor amongst all these domains is the task to solve a major complex problem that independent and like-minded contributors that include commercial partners and associates offer their asssitance through segmented versions of their solution in an accumulative attempt to get to a solution. Contributors share their experience and expertise as they connect, communicate, collaborate and collectively learn, the four foundation keystones of crowdsourcing [35].

The application of crowdsourcing to education is not so popular or practiced but had been initially proposed by Weld et al., [36] in 2012 and which was adopted two years later in a project to investigate how to enhance e-learning effectiveness [37]. This project that was finalised in 2016 coupled crowdsourcing with personalisation techniques and employed numerous social media to collate expertise and supplementary content for the previously set education material as part of an e-learning course. Crowdsourcing can also be evidenced indirectly within MOOCs as the collective feedback and collaborative authoring mentioned earlier is a joint effort by the numerous learners following the same course. Similarly the repository of educational resources made available by numerous educators and content providers is again a form of crowdsourcing that can be fruitfully and productively employed to source and supplement e-learning courses. Much of such processes could easily be automated through purposely developed software that will harness the power of the crowds to the advantage of e-learning. Similarly crowdsourcing concepts can be constructively employed to provide assessment and assistance to the same students while following an e-learning course. More about this in Chap. 5.

3.4 Social Networks

Social media have been on the rise since their inception in the late 90s with an outstanding popularisation that shook the world. User generated content and the force of masses can be considered an amazing combination that are amazingly overwhelming as they are grounded within human nature itself. As social beings we humans in general are enticed by egocentric tendencies to show off yet with a societal magnetism to share and collectively contribute. Social networks reinforced this dynamism as virtual communities, discussion groups, available for a enabled the individual to be counted and given the much desired limelight.

Numerous social media applications have been dominating peoples life as ubiquitous mobile applications have pervaded all aspects of our lives. Sharing, contributing, commenting, posting, tagging, discussing and prying on others activities is what a lot of online users do all day long. Social networks have filled a void within society and the assortment of associated media have taken full advantage while addressing a variety of aspects of our lives, one of which is education. Junco et al., [38] claim that Tweets with a learning environment increased engagement levels with both the learners and the educators as students participated more during activities and teachers were mobilised into a role which much more dynamic and engaging. Another study [39] confirmed the strong correlation between the use of social media and the positive vibes students experience during the learning process. These encouraging and optimistic sensations were also confirmed by numerous researchers [40–42] who highlighted the fact that incorporating social media as part of the learning process was a positive step towards the employment of a natural communication channel for learners who found it much easier to collaborate and collectively learn.

The academic relevance of social networks is most evident as they manifest themselves within MOOCs that have been expanded earlier. The marketing of these e-learning platforms are advertised, liked and suggested by numerous friends and social media users attracting others in a recursive way. Additionally the use of social networking tools and capabilities are enabled in these and other e-learning platforms and programmes whereby participating citizens [43] have the opportunity to consume as well as contribute to the global knowledge in a crowdsourcing way to match particular educational contexts and requirements.

3.5   Educational Implications

The educational implications of the above sections are numerous and very significant to the future of e-learning as numerous educational researchers [44–47] have pointed out over the years the importance and relevance of developing the precise techniologies to ensure they deliver an elevated e-learning system that is far more effective and functional. In line with this fact of technology leading the e-learning evolution, Web 2.0 technologies have not only left an impact on our online experiences but have enabled new traits, habits and practices that accompany us within e-learning environments. Such procedures are expected to be included and they are not only popular and conducive to enhance participation, but also beneficial in addressing a number of e-learning concerns highlighted earlier in the chapter.

Social networks have been argued and sustained [48] that they reduce loneliness and isolation especially within groups of students and their educators as they get to know each others interests and opinions. Any participant can initiate a topical discussion, give feedback to others, and complement further content to sustain an argument or to make a point. Such situations give participants, especially the learners, a sense of belonging as well as academic security resulting in a potential improvement in performance.

Similarly, motivational levels have been reported [49] to have soared through the additional use of social networks due to their dynamic interactive nature and flexible availability that enables collaboration, bi-directional communication and media-rich interaction. Such media cultivated confidence in learners who were able to author and curate content as part of their educational experience, as well as fostered creative traits and encouraged participation and self-expression.

3.6   Conclusion

The contribution of technology to the enhancement of e-learning did not fail to fulfil its capacity and potential as it did since the inception of e-learning. We have witnessed how the latest technologies and developments within the same e-learning camp have maintained the same trend with a parallel evolution in an attempt to optimise the electronic medium as an educational conduit. The discrepancy or rather the added bonus from the usual shadowing trend involved the unintended addressing of numerous pedagogical issues. Such a positive consequence will lead the way to the next chapters as we approach the inception of future e-learning systems through the evolution of e-pedagogies rather then technologies, and the addressing of specific academic dimensions rather than blindly adopting the latest technology.

References

1. Bloom, B.: The 2-sigma problem: the search for methods of group instruction as effective as one-to-one tutoring. Educ. Res. **13**(6), 4–16 (1984)
2. Vidakovic, D., Bevis, J., Alexander, M.: Bloom's taxonomy in developing assessment items. J. Online Math. Appl. (2003)
3. Skylar, A.A., Higgins, K., Boone, R., Jones, P.: Distance education: an exploration of alternative methods and types of instructional media in teacher education. J. Special Educ. Technol. **20**(3), 25–34 (2005)
4. Kartha, C.: Learning business statistics: online vs traditional. Bus. Rev. **5**(1) (2006). Cambridge
5. Piccoli, G., Ahmad, R., Ives, B.: Web-based virtual learning environments: a research framework and a preliminary asssessment of effectiveness in basic IT skills training. MIS Q. **25**, 401–427 (2001)
6. Suanpang, P., Petocz, P.: E-learning in Thailand: an analysis and case study. Int. J. E-Learn. **5**(3), 415–439 (2006)
7. Halawia, L., McCarthy, R., Piresc, S.: An evaluation of e-learning on the basis of bloom's taxonomy: an exploratory study. J. Educ. Bus. **84**(6), 374–380 (2009)
8. AP: Research on the effectiveness of online learning - a compilation of research on online learning, Academic Partnerships, Texas, USA (2011)
9. Chan, A.Y.K., Chow, K.O., Jia, W.: A framework for evaluation of learning effectiveness in online courses. In: Zhou, W., et al. IXWL2003. LNCS, vol. 2783, pp. 383–395. Springer, Berlin (2003)
10. Bernard, R., Abrami, P., Lou, Y., Borokhovski, E., Wade, A., Wozney, L. How does distance education compare to classroom instruction? a meta-analysis of the empirical literature. Rev. Educ. Res., pp. 379–439 (2004)
11. Means, B., Toyama, Y., Murphy, R., Bakia, M., Jones, K.: Evaluation of evidence-based practices in online learning: a meta-analysis and review of online learning studies. U.S. Department of Education, Office of Planning, Evaluation, and Policy Development, Washington, D.C. (2009)
12. Xu, D., Smith Jaggars, S. Online and Hybrid Course Enrollment and Performance in Washington State Community and Technical Colleges. Online Education and Instructional Technology (2011)
13. Johnson, H., Cuellar Mejia, M. Online Learning and Student Outcomes in Californias Community Colleges, California, USA, Public Policy - Institute of California (2014)
14. Neuhauser, C.: Learning style and effectiveness of online and face-to-face instruction. Am. J. Distance Educ. **16**(2), 99–113 (2002)
15. ODonoghue, J., Singh, G., Green, C.: A comparison of the advantages and disadvantages of IT based education and the implications upon students. Interact. Educ. Multimedia, **9**, 63–76 (2004)
16. Olson, J., Codde, J., deMaagd, K., Tarkleson, E., Sinclair, J., Yook, S.: An Analysis of e-Learning Impacts and Best Practices in Developing Countries. Michigan State University, Michigan (2011)

17. Noesgaard, S., Ørngreen, R.: The effectiveness of e-learning: an explorative and integrative review of the definitions, methodologies and factors that promote e-learning effectiveness. Electron. J. eLearning, **13**(4), 278–290 (2015)
18. Tang, T., McCalla, G.: Beyond learners interest: personalized paper recommendation based on their pedagogical features for an e-learning system. Lect. Notes Comput. Sci. **3157**, 301–310 (2004)
19. Weller, M.: The distance from isolation: why communities are the logical conclusion in e-learning. Comput. Educ. **49**, 148159 (2007)
20. Bousaaid, M., Ayaou, T., Afdel, K., Estraillier, P.: System interactive cyber presence for e-learning to break down learner isolation. Int. J. Comput. Appl. **111**(16), 975–8887 (2015)
21. Davies, J., Merchant, G.: Web 2.0 for Schools - Learning and Social Participation. Peter Lang publishers, New York (2009)
22. Siemens, G., Downes, S.: Current/future state of higher education, edfuture. (2012). http://edfuture.net/ Cited 12 Feb 2013
23. Coursera: Online Courses from Top Universities. (2011). https://www.coursera.org/ Cited 12 May 2017
24. Udacity: Personal learning environments. In: Proceedings of the Sixth IEEE International Conference on Advanced Learning Technologies. IEEE Computer Society, Washington, DC, pp. 815–816 (2011)
25. Cox, L.: Truth in crowdsourcing. IEEE J. Secur. Priv. 74–76 (2011)
26. Ramakrishnan, A., Halevy, A.: Crowdsourcing systems on the World-Wide Web. Commun. ACM **54**(4), 86–95 (2011)
27. Bernstein, M., Brandt, J., Miller, R., Karger, D.: Crowds in two seconds: enabling realtime crowd-powered interfaces. UIST (2011)
28. Bernstein, M., Little, G., Miller, R., Hartmann, B., Ackerman, M., Karger, D.: Soylent: a word processor with a crowd inside. UIST (2010)
29. Casal, D.P.: Crowdsourcing the corpus: using collective intelligence as a method for composition. Leonardo Music J., 25–28 (2011)
30. Belleamme, P., Lambert, T., Schwienbacher, A.: Crowdfunding: tapping the right crowd. In: The International Conference of the French Finance Association, AFFI (2010)
31. Harvey, D., Kitching, T.D., Noah-Vanhoucke, J., Hamner, B., Salimans, T., Pires, A.M.: Observing dark worlds: a crowdsourcing experiment for dark matter mapping. Astron. Comput. **5**, 35–44 (2014)
32. Fitt, V.: Crowdsourcing the news: news organizaion liability for ireporters. William Mitchell Law Rev. 1839–1867 (2011)
33. Bommert, B.: Collaborative innovation in the public sector. Int. Public Manag. Rev. 15–33 (2010)
34. Cosley, D., Frankowski, D., Terveen, L., Riedl, J.: Suggestbot: using intelligent task routing to help people and work in wikipedia. In: Conference on Intelligent User Interfaces (2007)
35. Literat, I.: The work of art in the age of mediated participation: crowdsourced art and collective creativity. Int. J. Commun. 2962–2984 (2012)
36. Weld, D.S. et al.: Personalised Online education - a crowdsourcing challenge. In: AAAI Workshops at Twenty-Sixth AAAI Conference on AI. [S.l.]: AAAI, pp. 159–163 (2012)
37. Montebello, M.: Investigating Crowd Sourcing in Higher Education. University of Sheffield. Higher Education in the Globalised Age, Sheffield (2014)
38. Junco, R., Heiberger, G., Loken, E.: The effect of Twitter on college student engagement and grades. J. Comput. Assist. Learn. (2010)
39. Rutherford, C.: Using online social media to support preservice student engagement. MERLOT J. Online Learn. Teach. **6**(4), 703–711 (2010)
40. Junco, R.: The relationship between frequency of Facebook use, participation in Facebook activities, and student engagement. Comput. Educ. 162–171 (2011)
41. McLeod-Grant, H., Bellows, L.: Leveraging Social Networks for Student Engagement. Monitor Institute, San Francisco (2012)

42. Churcher, K., Downs, E., Tewksbury, D.: Friending vygotsky: a social constructivist pedagogy of knowledge building through classroom social media use. J. Eff. Teach. **14**(1), 33–50 (2014)
43. Jenkins, H.: Confronting the Challenges of Participatory Culture: Media Education for the 21st Century. MIT Press, Cambridge (2009)
44. Manouselis, N., Sampson, D.: Dynamic knowledge route selection for personalised learning environments using multiple criteria. In: Proceedings of Intelligence and Technology in Education Applications Workshop (ITEA 2002), International Symposium on Artificial Intelligence and Applications, 20th IASTED International Conference on Applied Informatics (AI2002), IASTED, Innsbruck, Austria (2002)
45. Dagger, D., Wade, V., Conlan, O.T.: Towards anytime, anywhere learning: the role and realization of dynamic terminal personalization in adaptive e-learning. In: Proceedings of Ed-Media 2003, World Conference on Educational Multimedia, Hypermedia and Telecommunications. Ed-Media (2003)
46. Conlan, O., Wade, V.: Evaluation of APeLSAn adaptive e-learning service based on the multi-model, metadata-driven approach. In: Adaptive Hypermedia and Adaptive Web-Based Systems, pp. 504–518 (2004)
47. Van Harmelen, M.: Personal learning environments. In: Proceedings of the Sixth IEEE International Conference on Advanced Learning Technologies. IEEE Computer Society, Washington, DC, pp. 815–816 (2006)
48. Miller, D., et al.: How the World Changed Social Media. UCL Press, London (2016)
49. Rosli, M.S. et al.: E-Learning and social media motivation factor model. Int. Educ. Stud. **9**(1) (2016)

Chapter 4
User Profiling and Personalisation

The shoe that fits one person pinches another;
there is no recipe for living
that suits all cases.

Carl Jung

Abstract Personalisation, user profiling and the use of machine learning techniques from the computer science arena fall under the umbrella of Artificial Intelligence or AI. Rather then going through all the technical details of machine learning and AI we will be looking into the conceptual application of such techniques, as well as the educational undertones of doing so. Personalisation features as a main component in this chapter due to its exceptional and remarkable property of improving a service or a product. We shall be looking into how such a widely employed technique in industry can be similarly applied to education that promises to alleviate and add-value to e-learning as we know them. The main concept behind such a technique is the capturing and representation of the specific user model or profile. This user representation is a living model that evolves over time and requires constant updating to ensure the profile realistically embodies the user or the learner in our case. As we shall investigate in the next sections the user profile is generally generated and trained using the user patterns and trends but also the interests, needs and choices that all indicate something specific about the user in isolation as well as in combination together. In another section we will also take an in-depth analysis of how user profiling can be optimised in the case of education in a similar attempt to encapsulate the specific and characteristic learner profile. We close this chapter with a look at recommender systems and how all the different parts mentioned above come together to the cause of enhancing education and the e-learning medium.

4.1 Commending Personalisation

The act of personalisation in itself is a process that adds value to whatever product or service it is associated with and applied to. Postma and Brokke [1] took it unto themselves to specifically investigate and prove the effects of personalisation

© Springer International Publishing AG 2018

M. Montebello, *AI Injected e-Learning*, Studies in Computational Intelligence 745,
https://doi.org/10.1007/978-3-319-67928-0_4

as they focus on the advantages and benefits of a personalised mode in contrast to a more generic one. The authors wanted to prove what logically already made sense to them and to others but empirically established and confirmed that personalisation has a profound potency that intensifies as content is individually targeted to each specific user triggering an overwhelming and above-average responsive interest. This comes as no surprise as people, out of their nature, genuinely desire and unconsciously entreat personalisation at a physical and psychological level. A personalised experience gives a person a sense of uniqueness and of being distinctively special that is atypical of the rest of the world. This sense of ascendancy and control distinguishes the single user from the crowd whereby the service or product being provided is explicitly tailored to that same user or learner. The fact that a learner is being exposed to academic content that has been compiled exclusively to match his or her preferences, choices and interests then the entire experience motivates and prompts the learner to do better and push further. Such a sensation is provoked out of the conviction that a special treatment is being delivered differently from what all the rest are being provided. The learner is implicitly controlling the content and the experience as it has been shaped, modelled and compiled upon the personal choices of the same learner. Psychologists like Johnson [2] associate such experiences with a healthy psyche and a positive, productive attitude, which augurs well to an academic setting. Such a simple yet powerful concept has been applied to a variety of domains within society, but probably the sphere where personalisation has been most commonly employed and taken full advantage of is within the sales and advertising industry. The simple fact of a salesperson recalling the clients names is testament of how to make a sale and increase profits. It has been scientifically proven [3] that a persons brain is activated in a unique way when hearing his or her own name triggering emotional ripple effects in numerous parts of the brains left hemisphere associated with social behaviour, long-term memory, as well as auditory and visual processing. Such an activity within the brain sheds light on the ramifications of a person hearing ones first name giving a warm special sensation and of being in control in one way or another.

Sales departments have long realised that empowering customers by giving them the virtual control of a situation and addressing the exact needs of each unique customer increases their chances of success. Advertisers are fully aware that if they address the right customers employing the correct medium and the respective information then they raise the probability of an effective and fruitful advertising campaign. Parallels can be drawn employing similar examples in the political camp, tourism industry, automobile, insurance and several others. Amazon [4] was one of the first companies to employ market customisation as it employs data-mining techniques to generate a customer profile and eventually recommend fitting products and services. Amazon also applies its personalisation algorithms to online retailers who preferred such a technology over other modes of advertising for a greater return-on-investment. Similarly, Netflix [5] is the second most successful commercial company taking full advantage of personalisation as it employs a strong and successful content recommendation engine to highlight specific content for its registered viewers.

Another reason why personalisation is commendable is because it manages to condense and moderate the arduos task to process the masses of information we encounter, better known as information overload [6]. Such circumstances happen when a person is overwhelmed by the amount of information made available and the very act of personalisation tends to, at least gives the illusion or perception, that only that information that is relevant is being presented and needs to be consumed. The simple act of receiving a personally addressed e-mail rather than a generic blanket greeting is enough for a recipient to ignore or for an e-mail client filter to categorise as spam. The use of artificial intelligence to subdue information overload has been successfully applied to search technology [7] as search engines in the 1990s had already been struggling with the onslaught of online content while users attempting to assimilate an excess of information.

In the following chapters personalisation will be applied to education as it is argued that its effectivity can also enhance the educational process. In reality if an educator within a classroom situation had to employ a methodology personalised precisely for every learner then the outcome would be much effective but not so easy to perform. This has been confirmed by numerous educational institutions [8] internationally that different learners have different needs and learning styles which renders inflexible methodology inadequate. To this effect personalised learning has been defined [9] whereby educators can extract technological benefits through the measurement, collection, analysis and report of data about learners and their contexts, for purposes of understanding and optimizing learning and the environments in which it occurs? (p. 4). The techniques employed towards the personalisation process is still similar across domains but the sensitivity of the outcome within an academic environment is far more sensitive than suggesting web pages or recommending items to purchase. Still the use of data tools and intelligent techniques will be required making use of learner specific data in an effort to enhance the effectivity of learning process through personalisation. This was also investigated by Siemens [10], who earlier conceptualised the connectivism learning theory, and argued that it was possible to achieve such personalisation in education through the connections of provided online information in a contextual sensible way with help of networking capabilites that give meaning to the connected sources. A notorious publishing house and a research institution [11] have come together to inject personalisation techniques in e-learning courses in form of customised feedback and personal educational guidance that was established upon the learners previous performances. The initiative was launched in 2013 and serviced over 4,000 university first years. In a similar initiative [12] another publishing house and the University of Edinburgh deployed a personalised interface for learners while at the same time the educators were able to adjust their academic material to suit the evolving course. The same publishing house struck similar collaborations [13] with two other universities, Arizona State and Colorado Boulder, whereby personalisation featured as the main functionality to deliver formative recommendation to specific learners based on their interaction with the tailored content. The CogBooks publishing house involved in these academic partnerships are very proud to be uniquely educating each and every student as the results obtained are encouraging [14]. Reddy [15] reports similar results in a learner-based system called

U-Pace at the university of Wisconsin-Milwaukee whereby learners were individually coached with tailored advice on their progress and personalised plan of action following their academic outcome. Massachusetts Institute of Technology offered free e-learning programmes through its MITx programme to experiment with personalisation of content to those students who explicitly declared their personal interests and academic needs. Another initiative was by the IMS Global Learning consortium that proposed a set metrics, Caliper [17], to define a student profile in a standard way across three hundred education institutions. By employing a common protocol the learner could access personalised content across the universities. The Bill and Melinda Gates foundation also contributed to personalise education as a grant scheme was set up, the adaptive learning Market Acceleration Program (ALMAP), together with an academic programme, Enlearn, to convert the classroom into an adaptive educational ambient whereby customised content and teaching would facilitate the learning process for each individual learner. Worth mentioning as well the iClass initiative [18] that was developed in Malasia to offer a personalised learning experience. This web-based project integrated numerous techniques to tailor the interface, the content and the environment as well as enabled the appending of additional foreign applications according to the needs and requests of the learner. iClass was successfully adapted by Oxford University and used within secondary schools to assist students but also to support educators through the instant performance analysis facility [19].

4.2   Capturing User Interests

Intelligent systems simulate human behaviour, or better still they perform actions that normally require a person to do, and this is what makes them special and smart. As humans we show how much we care about each other or the way we show affection to one another is usually by paying attention to detail and to specific aspects that are unique to that particular person. The process of actually apprehending and attaining the finer details of a personality in itself is a skill that not everyone possess or capable to exert. Like any other skill, paying attention to detail and figuring out what the specific singularities that collectively distinct one person from another, can be picked up, acquired and fruitfully employed. The eventual utilisation of such a skill in reality, especially the positive or negative feedback that follows and the consequence of giving a special treatment to someone, will either reinforce or diminish and even erase the specific details that had accumulated about the same person. This cyclic process that we sometimes engage in, especially persons within some kind of relationship, is characteristic to humans and requires a degree of intelligence and common sense. Can we employ software algorithms to simulate this human skill and behaviour to capture the users interests? Explicitly asking the user directly to identify his or her interests involves no intelligence at all, and in fact numerous systems employ such a simple yet effective methodology of asking the user directly to specify and declare interests. There are a number of issues associated with this mode of capturing user interests. First and foremost such a technique is usually employed at the very beginning of a process, either when a user registers for the first time on a portal or when a novel system is being installed or accessed. This is performed to ensure that the functionality of the portal or system triggers in right from the start offering recommendations or suggestions to the user without going through what is commonly referred to as the cold start. The cold start problem [20] refers to the inability of intelligent systems programmed to recommend and suggest relevant information or media to users to do so when they have no data or history to base or drive their algorithms on. As a result one way to overcome such an inconvenience or inability to generate initial recommendations or suggestions the explicitly stated interests by the user are commonly employed to either power the algorithms directly, which could be too generic, or attempt to assimilate the user within a cluster of similar users like for example employing techniques like Collaborative Filetering [21] or the Social Choice theory [22]. The obvious drawback with this modus operandi boils down to human nature, as people registering for the first time or installing a new system would provide minimal input and bare compulsory details to get going. This setback jeopardises the appropriate functioning of any technique that attempts to address the cold start problem.

Another significant issue related to the explicit specification of user interests to be captured by the underlying software system to personalise recommendations and customise suggestions is the factual reality that such interests vary and rapidly evolve over time. So unless the intelligent recommender is designed and intended to employ other variables like implicit capture of user interests as time goes by, then the system is bound to fail and generate out-of-date proposals and irrelevant advice. A possible work-around both issues is to initially capture explicit user interests to overcome the cold start keeping in mind that issues related to impatient users could still persist, while enabling some kind of dynamic system to implicitly capture user interests as time goes by and the cold start problem subsides. Numerous attempts [23–26] have been documented that portray the combination of different methodologies while employing a variety of techniques. Results varied across the studies especially as the domain of application varied from news to movies, shopping, educational material and multimedia content. This goes out to show how fluid and unstable this research area is as the human factor will always remain an unknown and uncontrolled variable.

Capturing the user interests and ensuring that such interests truly characterise and realistically embody the user as accurately as possible, is only the initiation of a much complex process. Once these interests are captured they can be employed to match and identify additional news content, personal items, multimedia, etc. that can be consumed by the particular user. To periodically perform such a task it is much more efficient and convenient to generate and maintain a specific user profile that uniquely represents the interests and priorities of each individual user. Additionally such a profile could effectively evolve and maintain a much more realistic user representation rather than employing the interests in isolation. How can this be applied to the academic scene and how could a learners educational record be encapsulated to be fruitfully employed within a recommender system?

4.3   Learner Profiling

The concept of permanently capturing the users interests within a dynamic and evolvable representative profile conveniently allows the reuse and continuous tweaking of that same profile. Within academic circles the learner profile captures and condenses all the required academic knowledge an educator would need to comprehend before teaching, counselling or instructing a particular student. The profile could take the form of numerous academic artefacts and achievements as well as additional information that summarises the learners specific skills, interests, preferences, strong points, as well as weaknesses, learning difficulties, and areas that require special attention. Such information in isolation but also as a composite corpus of facts about the learner will help customise the content and delivery of an ordinary e-learning course into a tailored course that addresses issues and highlights those matters that need attention more than others. The learner profile in itself can be employed for a number reasons apart from assisting the personalisation process. Clustering same-minded learners could be beneficial within a physical classroom while facilitating the social aspect between such students within an e-learning setting could also be possible through the analysis of their profile. Numerous entities can benefit from a learner profile and not just an intelligent automated system what will employ it as training data to its AI algorithms.

The learner would be interested to explore and inspect his or her academic profile as it takes shape and evolves, while will feel inclined to improve, curate and perfect it. It is a well-known phenomenon that people tend to perform better when envisaging a goal or when specific intermediate targets are set. Research [27] has shown that, similar to intermediary goal attainment in games and use of achievement badges, students engagement and motivational levels improve as these additional pedagogical techniques are perceived as cognitively rewarding with optional goals, challenges and achievements that are publicly visible positively affecting the students behaviour.

Educators also benefit from learner profiling as they can effectively adapt their teaching methodologies to the different needs of the learners. In reality it is quite challenging to plan and prepare a differentiated lesson to address all the different needs of the diverse learners profile. The fact remains that once an educator is fully aware and apprehends what the needs of the specific learner are as a result of the explicit and subjective academic profile then a conscious and intentional effort to assist, service and support the educational needs of the learner can be done. From a pragmatic point of view an educator can cleverly and strategically devise a lesson plan and employ a variety of teaching approaches that allow the possibility for individual students or groups of students to adopt one particular stance that appeals to their learning approach rather than other. This way to differentiate between teaching strategies that suits the heterogeneous needs of the different learner profiles requires extensive planning, informed pedagogical choices and decisions, as well as a capacity to flexibly adapt and execute a variety of instructional undertakings. This is precisely what an intelligent e-learning system needs to aspire to. Not just assisting an educator performing a single yet important task, but facilitating that same educator to excel and optimally perform when faced with a challenging and intricate situation of students with a plethora of learning profiles. Such help could take the form of content selection, learning preferences, teaching aids employed, examples relevant to the learners interests, and assessment methods that best work with the specific individual.

To what extent can we trust a learner profile that is being curated by a student? The contents as well as the frequency to update the learner profile are instrumental in its validity, effectiveness and purpose together with the overall success of its utilisation. It is precisely for this reason that an automated and intelligent system that performs such a task is much more reliable, secure, and trustworthy as it takes into consideration every academic aspect without any prejudice, bias or self-regard. When an intelligent automated system generates a learner profile as a digital representation of the academic depiction of a students achievements and experiences, which is commonly referred to as a personal learning portfolio. We will be looking into how Personal Learning Portfolios (PLP) have evolved in the next chapter and how they truly represent a specific student and how they can be fruitfully and conveniently be employed to train, generate and maintain up-to-date the learner profile.

4.4 Educational Recommender

The reason behind employing a learner profile is to optimally assist learner in their academic experience. Educators would be in a better and advantageous position if they were aware and fully knowledgeable of the complete educational history of their learners. They would be capable to appreciate even better the exact educational needs and requirements, as well as the most appropriate teaching methodologies and optimal pedagogical settings to confer and deliver the ideal and most adequate personalised teaching settings. Such settings would include the medium employed, techniques adopted, subject domain, pitching level, assessment methods, pace of delivery, and all other minute factors that inevitably affect the student and the learning process. Using the same quintessence parallelism described above to an automated and intelligent system that promises to similarly deliver an ideal learning environment through the engagement of the academic profile that is specific to each unique learner. Such a system is not intended to replace the educator in any way but simply to provide smart educational recommendations while at the same time automatically update and evolve the same PLP in the process.

Educational recommenders are not a new concept and have been around since the 1960s and 70s in some rudimentary form when AI was still in its infancy. Since then they have evolved immensely and worth mentioning an excellent effort to provide personalised academic recommendations through systems that are termed as adaptive educational hypermedia. These systems have mainly four parts [28] that collectively enable the tweaking of the content that is the first required element of any recommender system. The knowledge base or space is usually populated either through educational resources provided by educators or through a private or public repository. Another component is the recommender algorithm itself that consumes the stored content to deliver customised material. Such an adaptation mechanism requires the learner profile that has been reviewed in detail in the previous sections and which constitutes the third vital component. Finally, as mentioned earlier, this user model is required to evolve over time and thereby some way of capturing the learners feedback, actions and outcomes is required. These observations are the training and sustaining material for the user profile or model to be generated and sustained up-to-date.

The administrator of such recommendation systems is required to ensure that system is performing as required and therefore needs to perform a number of tasks in reality require dedicated professionals. Any educational system, even the traditional face-to-face, requires the knowledgeable person or content expert that is required to specify the content and basic structure of the academic material. A recommender system could potential employ an all-encompassing subject syllabus within its knowledge base that the adaptation mechanism will select from. However some structure within the content still needs to be maintained and thereby a design is required to be followed that would have been encoded with the system following some pedagogic methodology set by corroborating academics. Finally, an online or electronic recommender system also requires well-designed interfaces that in themselves could be intelligently put together. Intelligent user interfaces is a research discipline in its own right but generic interfaces designs can be set by professional designers to ensure that the academic content and the pedagogical methods employed are presented in the best way possible that accommodates the particular learner. This brings into perspective that the whole process has a precise and sequential order that needs to be followed. Brusilovsky [29] identifies four key steps that are involved within an adaptive hypermedia recommender system, namely:

1. Employing a structured domain model at the centre of the recommender system to successful target and effectively deliver the correct content from the efficiently designed knowledge base. The semantic web has much to offer on this component as the use of ontologies simplifies and automates the meaningful capture, storage and use of content especially the copious and heterogeneous education material that is freely available online. This step does not include the content itself but the structure that will sustain and support it in a rational and consistent way.
2. Adopting a personal academic profile methodology to capture and conserve the learner model in a way as described in Sect. 4.4 to ensure a realistic and true learner representation. Use of ontologies to represent such a model can be matched to the previously described domain model to identify overlaps and thereby detecting what the learner might be interested in and what will fit the specific learning profile. The PLP is also required to adapt and evolve over time which makes ontologies ideal due to their dynamic and easily adaptable nature.
3. Designing the educational medium space that will be used to describe each and every resource to be employed. In line with the two previous steps the use of technologies like RDF (Resource Description Framework) will make it easier to integrate and fruitfully employ any of these resources with the other ontological

aspects. A robust and comprehensive methodology to describe and capture all the attributes of a particular education resource, including academic relevance and associations, will support and improve the successful engagement and effective utilisation of the resource itself. A strong and secure educational medium space the greater the chances of a successful and productive recommendation system.

4. Applying adaptation rules to match the learner model to the domain model and the medium space in a way to be able to identify, retrieve and employ specifically selected education resources as part of an academic programme. The algorithmic rules will ensure that the learners interests, characteristics, needs and preferences, through the PLP, will be catered for and fully addressed when selecting the academic course to follow, the resources employed, and the methodology adopted.

The methodologies employed and the techniques adopted are subject to continuous discussion as researchers have hypothesised, tested and evaluated numerous approaches conducted in a variety of ways. Some methods perform better under certain circumstances and particular domains, while others accomplish superior recommendations when employed in different situations and exposed to diverse conditions. The truth being that there is no one magical combination that outperforms all others in every situation and applied to all domains.

4.5 Conclusion

In this chapter we have started to browse into how AI will play an important role in the evolution of e-learning. The key concept that renders all the work, effort and investigation worth it is the phenomenon behind personalisation. It was argued that personalisation is an intensive and compelling element as it factors in with ingrained human characteristics, which can be taken advantage of to assist and deliver an enhanced e-learning experience. To do so a number of contributing issues were discussed and put into perspective within an educational recommender. The use of ontological representations was also discussed as an ideal structure to model the domain knowledge base, as well as the user interests that need to be captured. The learning profile was discussed in some detail in order to encapsulate the specific and unique academic traits of each student, which evolves as the learners feedback and interaction with the recommender further fuels the adaptation AI techniques in the background. How will all this come together? The next chapter proposes a model that complies with such a setup as the goal to customise e-learning draws closer.

References

1. Postma, O., Brokke, M.: Personalisation in practice: the proven effects of personalisation. J. Database Mark. **9**(2), 137142 (2002)
2. Johnson, J.A.: Freedom and Control. Psychology Today, Sussex, 30 (2011). https://www. psychologytoday.com/blog/cui-bono/201104/freedom-and-control. Cited 21 January 2015
3. Carmody, D.P., Lewis, M.: Brain activation when hearing ones own and others names. Brain Res. **1116**(1), 153158 (2006)
4. Linden, G., Smith, B., York, J.: Amazon.com Recommendations Item-to-Item Collaborative Filtering, Internet Computing, pp. 76–80 (2003)
5. Gomez-Uribe, C.A., Hunt, N.: The netflix recommender system: algorithms, business value, and innovation. Trans. Manag. Inf. Syst. **6**(4), 1–15 (2015)
6. Melinat, P., Kreuzkam, T., Stamer, D.: Information overload: a systematic literature review. In: Johansson, B., Andersson, B., Holmberg, N. (eds.) Perspectives in Business Informatics Research. [S.l.], pp. 72–86. Springer (2014)
7. Montebello, M.: Metasearch + Machine Learning = WWW Information Overload. J. Comput. Inf. (JCI), Canada, **3**(1) (1999). ISSN 1201-8511
8. Educause: Personalized Learning (2016). https://library.educause.edu/topics/teaching-and-learning/personalized-learning. Cited 15 September 2016
9. Lonn, S., Nixon, A., Morgan, G., VanDenBlink, C., Dahlstrom, E.: Moving the Red Queen Forward: Maturing Learning Analytics Practices, Educause15 (2015)
10. Siemens, G.: SenseMaking Artefacts, Connectivism (2012). www.connectivism.ca. Cited 12 Nov 2015
11. Knewton: Pearson and Knewton Team Up to Personalize Math Education, Knewton in the News (2016). https://www.knewton.com/resources/press/pearson-and-knewton-team-up-to-personalize-math-education/. Cited 20 Dec 2016
12. CogBooks: Using adaptive learning tools An educators perspective (2015). https://www.cogbooks.com/2015/09/15/using-adaptive-learning-tools-an-educators-perspective/. Cited 18 Nov 2016
13. Lawlor, O.: Metacog releases open ended rubric based machine scoring service (2015). http://www.metacog.com/blog/files/category-assessment.html. Cited 30 June 2016
14. CogBooks: Improve Student Success and Retention with Adaptive Courseware (2016). https://www.cogbooks.com/2016/02/04/improve-student-success-and-retention-with-adaptive-courseware/. Cited 12 Dec 2016
15. Reddy, D.M.: U-Pace. University of Wisconsin-Milwaukee. Educase (2014)
16. MIT: MITx, Free online courses from MIT (2016). https://www.edx.org/school/mitx. Cited 8 Nov 2016
17. IMS: Learning Measurement for Analytics Whitepaper, IMS Global Learning Consortium, Inc. (2013). https://www.imsglobal.org/sites/default/files/caliper/IMSLearningAnalyticsWP. pdf. Cited 15 Sept 2016
18. Ined: Institute of Neo Education, iClass Learning Management System (2016). http://iclass. ined.uitm.edu.my/. Cited 9 Feb 2017
19. OUP: Oxford Universtiy Press (2016). https://www.oupchina.com.hk/elt/events/20160305-iclass-seminar. Cited 12 Dec 2016
20. Bykau, S., Koutrika, G., Velegrakis, Y.: Coping with the Persistent Cold-start Problem. In: Personalized Access, Profile Management, and Context Awareness in Database, PersDB 2013 (2013)
21. Takacs, G., Pilaszy, I., Nemeth, B., Tikk, D.: Scalable collaborative ltering approaches for large recommender systems. J. Mach. Learn. Res. **10**, 623–656 (2009)
22. Li L., Tang X.: A solution to the cold-start problem in recommender systems based on social choice theory. In: Lavangnananda K., Phon-Amnuaisuk S., Engchuan W., Chan J. (eds.) Intelligent and Evolutionary Systems. Proceedings in Adaptation, Learning and Optimization, vol. 5. Springer, Cham (2016)

23. Li, L., Zheng, L., Yang, F., Li, T.: Modeling and broadening temporal user interest in personalized news recommendation. Expert Syst. Appl. **41**(7), 31683177 (2014)
24. Burke, R.: Hybrid systems for personalized recommendations. Intelligent Techniques for Web Personalization, p. 133152 (2005)
25. Chu, W., Park, S.: Personalized recommendation on dynamic content using predictive bilinear models. In: Proceedings of the 18th international conference on world wide web, p. 691700. ACM (2009)
26. Albanese, M., Chianese, A., d'Acierno, A., Moscato, V., Picariello, A.: A multimedia recommender integrating object features and user behavior. Multimed. Tools Appl. **50**(3), 563585 (2010)
27. Hakulinen, L., Auvinen, T., Korhonen, A.: The effect of achievement badges on students behavior: an empirical study in a university-level computer science course. Int. J. Emerg. Technol. Learn. **10**(1), 18–29 (2015)
28. Henze, N., Nejdl, W.: A logical characterization of adaptive educational hypermedia. New Rev. Hypermedia Multimed. (NRHM) **10**(1), 77–113 (2004)
29. Brusilovsky, P.: Developing adaptive educational hypermedia systems: From design models to authoring tools. In: Murray, T., Blessing, S., Ainsworth, S. (eds.) Authoring Tools for Advanced Technology Learning Environment, pp. 377–409. Kluwer Academic Publishers, Dordrecht (2003)

Chapter 5
Personal Learning Networks, Portfolios and Environments

*Always remember that
you are absolutely unique.
Just like everyone else.*

Margaret Mead

Abstract The initial steps towards the model to personalising e-learning through the injection of AI starts to take shape in this chapter as several of the factoring elements have been covered in the previous two chapters. These will form part of a personal learning environment that each individual learner or life-long learner would establish and assemble around oneself in an effort to create and avail of a sustainable educational system that has the learner at its centre. Personal learning environments or PLEs are ideal vessels to encapsulate all a learner requires due to their personalisation capabilities that truly empower the same learner. Morrison [1] identifies two essential components within a PLE as he depicts its anatomy as shown in Fig. 5.1 overleaf. Each of these components play an important role and need to be investigated individually to ensure that they are optimally setup and compatibly designed to generate the expected outcome, an intelligent personal learning environment. The Personal Learning Network (PLN) and the Personal Learning Portfolio (PLP) form part of the PLE and will be initially presented in the following sections as they bring together essential components from the previous chapters. These underlying technologies that source both of them will be justified in terms of their academic relevance, pedagogical effectiveness, and theoretical suitability. How will both these components take advantage of the latest technological developments and boost well established technologies in an attempt to enrich the learning experience? How will the different technologies compatibly come together to ensure the learner is not just at the centre of the PLE but also in full control of the medium? The personal learning network and the personal learning portfolio are intended to complement each other as they form part of the proposed model in the next chapter.

© Springer International Publishing AG 2018
M. Montebello, *AI Injected e-Learning*, Studies in Computational Intelligence 745,
https://doi.org/10.1007/978-3-319-67928-0_5

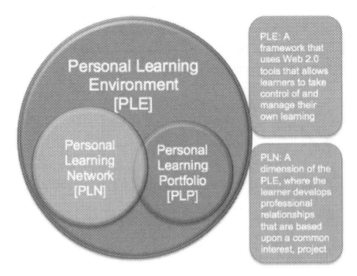

Fig. 5.1 Anatomy of a PLE [1]

5.1 Personal Learning Network (PLN)

The concept of a personal network is usually associated with a communication network that entails a physical network of devices around a human person, also referred to as a Personal Area Network (PAN). The devices that could potentially include a networked mobile phone together with camera, display glasses, storage device, wearables and other devices like headphones, watch, speaker, etc. The main reason behind a personal area network is for the user to enjoy the access to all the required input and output devices while connected to the web and all the online information. The PAN equips the user with all the required input sources and output devices to optimally communicate and operate while taking advantage of the wirelessly connected technologies. Employing the same concept of a personal network to a learning environment, in an ideal situation the learner is surrounded by a set of virtual educational resources to assist the same learner in all the personal academic needs. Such a support system would be highly effective and recommended as Leone [3] points out the necessity and significance of having a healthy and supportive system around a learner made up of available web resources and crowdsourced knowledge from social networks. Such a customised learning network is closely interwoven around the learner and continuously evolves in tandem with the learner as new sources are appended and as relationships mature and develop. The interaction itself between the learner and the associated learning network is also an agent of change and adaptation as educational processes develop, learner needs alter, and new resources become available over time. A personal learning network enables the learner to develop, curate and make good use of those resources and sources that adequately support and optimally assist each individual learner in a unique way. This was not always possible as students could only surround themselves with books and notes from their scholastic interactions with educators. However as the connected online presence pervasively and ubiquitously dominated the learners life it became much more realistic and practically possible to build such a personal learning network.

Education researchers eventually justified such a phenomenon as the learning theory of connectivism was proposed by Siemens [4] and defended by Downes [5]. Connectivism empowers the learners through their online connected capabilities as they employ them to gain access, interact and take academic advantage of the freely available resources and materials. The educational process can be found not just in the content and within knowledge-bases, but also in the myriad of possible activities that the interactive Web 2.0 enables and permits. Learners have the potential to tap into boundless resources that are not possible through classroom interaction. They can collaborate, interrelate and cooperate with other learners within their own personal learning network as they share, contribute, and generate ideas and information. Web 2.0 tools enable and impel students even more as they are encouraged and induced to participate as participation and belonging within a healthy and supportive learning network further induces and motivates education. Additionally a PLN reduces issues of isolation as learners are free and willing to boost and amplify their circle of academic contacts, resources and connections. This generates a novel way of life-long learning where the learner autonomously and independently chooses to pursue specific educational routes and particular learning networks in a self-determined manner. The learner freely elects to initiate new instructional interests, resume and regenerate previous ones, and investigate potential alternatives, as effective educational practices to search, identify, retrieve, collect, reshape, collate and curate enable effective educational processes. Tools that enable and facility such academic connections are widely employed [6] and effectively utilised [7] as learners dynamically contribute to the successful distribution of knowledge.

Highly relevant to populating the PLN is the concept of crowdsourcing that was introduced earlier in Sect. 3.3 whereby it is possible to merge together numerous online elements. These elements can potentially aggregate a plethora of skills, rich know-how, as well as the required expertise to provide a tailored educational experience customised to ones needs. The web has provided the platform for learners to share, cooperate and contribute from their free will to augment the already massive online knowledge base. Researchers [8] have successfull shown that such technologies are able to link knowledge workers with their respective online personal networks to exchange information within their supporting network and informal learning which substantiates the argument that PLNs add value as they take full advantage of the benefits extracted from social networks. PLNs have an important role in such a learning scenario and which will form an essential part of the model in the next chapter. However other essential components are required to complement the rich resource provider that the PLN provides.

5.2 Personal Learning Portfolio (PLP)

There have been numerous interpretations of what a learner profile refers to in contrast to what one intends when mentioning a personal learning portfolio or even an electronic portfolio or e-portfolio. The term personal learning portfolio, or PLP, here refers to the digital or electronic version of what a learner profile encapsulates. Baumgartner [9] distinctly subscribes to this definition as he refers to a PLP to be a digital representation that have classically featured within a learning profile of a student. In his taxonomy of such portfolios he identifies twelve different categories that all play a crucial role in a persons academic achievements and accomplishments. The personal learning portfolio should portray three aspects of a learners academic profile. First of all a PLP is required to furnish a proper reflection of the real academic experiences that the learner came across like the different courses, topics, content, assessments and curriculum. These could potentially be retained by the learner or the educational institution, where some of which are final products or achievements and others are academic processes, which in either case say something about the learner. A second aspect of the PLP should clearly indicate the development processes that the learner went through. These experiences can also be personal or institutional and products or processes, like for example job experiences, competences, qualifications, and other aspects that are normally stated within a curriculum vitae. Finally, a third PLP aspect should portray the best selling points of the learner represented in an optimal way to showcase the assortment of academic potentials as well as the educational aptitudes and competencies of the individual. A personal learning portfolio is required to be versatile and all encompassing as it is required to cater and address a variety of learners of different ages, backgrounds, cultures and nationalities. Additionally the same PLP will be compelled into action within different contexts and situations that will potential involve a array of diverse learning stages together with a immense combination of distinctive and blended information. Such requirements demand a PLP with a robust configuration that theoretically accommodates all possible scenarios of any type of learner. Personal learning portfolios can be advantageously employed within an online environment as access to the right resources to match the same portfolio is easier and much more efficient. This means that learners are much more inclined and prompted to take matters in their hands enabling a healthier student-centred philosophy as they take full control of their educational needs. Through the in-depth analysis of all three PLP aspects mentioned above it is also possible to integrate and support alternative ways of consuming new forms of academic content together with diverse and suitable assessment methods that are tailored to the specific learner. This provides the required stability, as life-long learners require as they switch between academic institutions, a variety of educators, and even diverse education systems around the world. The same can be said about those individuals who undergo continuous professional development as their needs and skills gap become even more elusive and subtle thereby requiring even more personalised attention. The PLP is the second element within the learning environment depicted in Fig. 5.1 and overlaps with the PLN as they share the learners individual characteristics that define and specify each and every one of us. In academia [10] a learning portfolio is usually defined as a students academic record that evolves over the years to accurately capture the work performed and the achievements attained. Lorenzo and Ittelson [11] affirm that such a definition is just one out of six categories of a PLPs functionality as it is employed to identify specific academic content that satisfies the tailored needs of the learner. Others [12] suggest that, in line with the self-determination learning theory, PLPs are practical tools that encourage self-directed learning that is reflective of the learners specific academic achievements. This alleviates motivational issues that numerous online learners experience [13], and which some researchers [14, 15] attribute the use of portfolios to a rise in enthusiasm as learners initiate and participate in additional learning processes especially within their academic network. It is precisely in this respect that PLP an PLN overlap as learners are able to share, collaborate and contribute within a learning network with the help of their portfolios [12]. This contributes to the escalation of individual learning processes, effective e-learning, and cognitive improvement [16]. This overlap between the PLP and PLN is also evidenced in other work [17] as student models materialise through the evolution of their PLPs and optimally employed to adapt and customise the entire learning environment. The authors reiterate the value of a PLP as it reflectively tailors future academic encounters to the specific learner and the respective learning interests, styles, and objectives. The way personal learning portfolios are generated has been discussed in the previous chapter and this does not involve the simple accumulation of learner artefacts and assessment results. The portfolio being discussed here is much more complex in nature and involves learning profiling to an extent that it is able to capture and represent the variety and extent of the learners interests and needs. Computer scientist have over the years developed numerous effective user profiling techniques as mention in Sect. 4.2 that are employed to generate a learner profile and utilised in the customised learning process. Such a learner profile is intended to comprehensively encompass characteristic information that is not necessarily observable. Such techniques have been widely employed [18] even within the simple browsing trends of a user to customise the browsing experience and reduce the information overload mentioned earlier. Similar techniques to generate and employ a PLP have been documented [19] whereby teaching materials were tailored according to the different profiles. Similarly Vargas-Vera and Lytras [20] made use of the e-learning environment itself to capture the learning trends of each learner and tailor future educational experiences of the same learner in an attempt to enhance learning process. Dagger et al., [21] had earlier shown that personalised learning through the customisation of learning environments and use of PLPs enhance the effectiveness of the medium employed. The consequence of such research [22] pushed the PLP as part and parcel of the PLN [23] and unfortunately focused on how to improve the e-learning application software rather than treating online education as completely different kind of web application that involved complex and intricate learner issues. The PLP needs to be investigated and developed as a separate entity that even though it overlaps with the PLN it should never become a sub component or an after-thought within the greater learning environment, the PLE, which will now be discussed.

5.3 Personal Learning Environment (PLE)

The personal learning environment or PLE brings together not just the previous two concepts, the PLN and PLP, but encapsulates the essence of personalised education. The combination of the personal network and portfolio are instrumental in establishing a healthy and fruitful learning environment as they assist in creating a close and personal academic atmosphere around the learner that ensure that every personal interest, need and educational requirement is taken into consideration. The notion of employing a PLE has been investigated by numerous researchers in a variety of ways. Conceptually it was formalised [24] as a learners personal assembly to support and contribute not just the normal learning process, but encourages new and different learning modalities as a variety of pervasive technologies and social networks have set the scene and made it all possible. Charlier et al., [25] attempt to define a PLE by distinguishing it from a stricter and resource-bound VLE (Virtual Learning Environment). The authors point out that the PLE is distinguishably learner centered especially in its use and application rather than having an educator dictating what resources are incorporated and which format and order the academic material is presented due to educator preferences, educational institution policy, and other restrictions.

5.3.1 PLE Categories

The PLE research area, even though in its infancy, has flourished alongside the WWW as educational researchers and commercial entities strived to design and develop an effective environment to enhance the learning experience. As a result of numerous research projects and technological developments different PLE models have emerged reflecting learning theories and respective epistemological reasoning. At one point three different categories were specified [26] to distinguish between PLE models, namely, base architecture, underlying platform, and pedagogical approach.

The first category distinguishes whether a client-server architecture is adopted for the PLE or whether a web-based infrastructure is deployed. The base architecture of the PLE will eventually determine numerous other factors that might influence its use, upgrade and eventual success. The client-server requires a client-side installation that communicates with the server to get serviced and access all resources. On the other hand the web-based alternative makes use of a web browser and thereby less demanding on the client as the communication aspect is catered by the browser itself. Any upgrades or novel functionalities to the PLE itself will not effect the web-based version while the client-server alternative will require a new download on the client side. Additionally the web-based PLE can easily adopt additional web services and resources that become available on a daily basis. In this respect Peter et al., [27] state that PLEs are an ad hoc, opportunistic aggregation of Web 2.0 services built to support a specific learning goal? (p. 1). ELGG [28] is a good example of employing atomic online services to provide an open-source platform that incorporates a social-networking engine. Socially-aware applications are hand picked by learners to create a personalised environment with social media and online resources. Another example of such a PLE is the PLEX personalised learning environment [29] makes extensive use of plugins provided by the same users who personalise the environment by integrating those components that most appeal to them and which they find useful.

The underlying platform that supports the learning environment is the second distinguishing factor to categorise PLEs. The norm for numerous established universities is to utilise and piggy-back their current learning platform by appending essential functionalities and capabilities to personalise the same environment according to their needs. In a study [30] to investigate learners preferences regarding the implementation of a PLE at the University of Southampton it was concluded that it is best to extend their current VLE and allow additional and supplementary components and services that include social media, communications and sharing facilities.

A final distinguishing element that PLE are able to be categorised is according to their pedagogical strategy that has been engaged. Three distinct instructional strategies that differentiate between types of PLEs are student-centered, institutionalised, and personalised, which will be expanded in more detail in the next section.

5.3.2 PLE Pedagogical Approaches

Pedagogy is the art or science of teaching and thereby it plays a crucial role in the blend of instructional strategies employed and that the designers and developers of the PLE set up and deployed. There exist numerous pedagogical approaches that are adopted in real life in classrooms and e-learning environments, but not all apply or are compatible with a PLE environment. One of the most liberal approaches that numerous PLEs adopt is the learner-centred approach where the student controls the entire activities. This approach has no specific strategy but entrusts the learner in initiating, controlling, selecting, and scheduling every academic event with the respective decisions that need to be taken and carried out. This obviously requires a good degree knowledge, self-control, discipline, and motivation to ensure that the self-regulated PLE provides a beneficial and advantageous learning environment whereby the academic process reflects good practice and informed decisions. The learner is able to compile the personal environment by picking, choosing and incorporating specific sources, services, and tools that s/he deems fit and which are compatible with her/his interests, needs and optimal way of learning. Numerous researchers [31] argue in favour of such an open approach where learners are free and self-determined to select and fruitfully employ educational tools, services and resources which suit them and which result in a much better and effective PLE. Ebner & Taraghi [32] also support this approach and positively conclude that learners from the technical university of Graz performed better when they liberally adapted the PLE with educational components of their choice. Such academic modules enable learners to tune the flexible environment exactly to their tastes and needs as they pick and plug web widgets to form part of their set of tools. Tools associated with search portals, social networks and multimedia provide the required resources to supplement the self-directed learning environment.

In the circumstances where the educator is allowed to contribute and get involved in the set up of the PLE then the second PLE approach comes into play. This approach goes beyond the learner in isolation and in full control, but envisages the educators and the institution itself in the overall management of the PLE. Garcia-Penalvo et al. [33] refer to this approach as the institutionalised PLE and together with other researchers [30, 34, 35] argue that this approach is academically safer as students are not left unaccompanied but educators are enabled to assist their students to adjust and personalise the learning environment to fit their agreed academic needs and interests. Such a semi-structured approach is seen by some researchers [36] as a perfect balance between student and teacher whereby the educational institution provides the platform with allowable services, tools and sources that the learner is encouraged, coached and guided to amalgamate any of the available educational items. The goal of a proposed framework [37] of a PLE that hosts this student-teacher approach is to urge and persuade educators as well as inspire and endow learners with the required motivation within a transformative cycle of creating PLEs that support self-regulated learning? (p. 6). To this effect the institutional learning environment needs to be open and compatible to auxiliary educations services and resources to enable an efficient and functional integration. So much so that some researchers [38] insist that at the rate of how freely available resources are materialising, it is important that institutional learning environments and VLEs are highly accessible and easily accommodating to such resources. This would render the same environments favourably useful and desirably inclusive in a way that productively supports learners in their academic needs by adding familiar and diverse tools and services.

The previous approach enabled personalisation or environment tweaking within confined limits as the institution provided a sort of sand box within which both the student and teacher are functionally constrained and academically restricted. The third and final approach distinguishes itself from the previous one as it enables complete freedom to both student and teacher to fully administer the PLE. Additional customisation functionality in this case goes beyond the aesthetics or the simple integration of available resources, but an unrestricted and unbound capacity to adopt any online resource, service or source of information by students and teachers. Researchers who approve and advocate this approach [39] regard a PLE as a learners personal environment which is molded according to the same learners education choices and decisions together with a distinctive self-managed academic direction.

5.3.3   Intelligent PLEs

A fourth and avant garde PLE approach is not just personalised and tailored to the needs and interests of the learner but smart and intelligent that takes personalisation to a conceptual level whereby it provides personalised pedagogical assistance to the learner such as recommendation of material, common interest learners, and adaptive path personal learning? as predicted by Al-Zoube [26], and continues by adding that future PLEs will enhance the quality of the instruction while reducing the demands of an instructional designer? (p. 60). Numerous PLEs attempt to implement an intelligent environment that would embody a one-to-one relationship between a student and a teacher, but the quality and effectiveness of such a PLE boils down to what the designers understand by intelligence and how it manifests itself. In one attempt /cite{Pearson5} an adaptable learning environment claimed to be intelligent assists the learner to customise the interface and the content according to the learners personal needs, academic requirements, and topical interests while involving additional contributing actors like knowledge experts, designers and developers apart from the traditional student and teacher. Such a PLE would easily fall under the third approach mentioned in the previous section, and once again raises the issue of what is intelligent. An intelligent PLE employs artificially intelligent techniques to offer services and functionalities that usually require a human to perform. In this case the intelligent component within the PLE is required to simulate, mimic and replicate as much as possible the essential and precious work of an educator. As the learner interacts with the PLE and produces activity data as well as academic outcomes the underlying intelligent PLE is expected, as usually provided by an educator, to adjust and further refine the learning environment to tailor it even more to the evolving learning profile of the student. To such ends a team of researchers [41] attempted to create an intelligent PLE and truly personalise the learning experience by employing portal technology. A portal is an all-inclusive web-based environment which provides the user with all that is required to get a task done including resources, tools and services. The intelligent component personalises the academic content based on information explicitly provided by the learner and presents tailored and meaningful educational material. To note that learners in this PLE were clustered according to their similar interests and that the underlying software did not extract any implicit information from the learners interaction with the same PLE. Another attempt [42] to create an intelligent integrates the facilities to plan and execute as part of the standard Learning Management System (LMS). In this way the PLE offers groups of learners an academic route as a plan was compiled by personalizing the content and tasks to their needs. The execution component keeps track of the performance of the proposed routes and adjusts the paths to optimise their performance on the next execution. The authors report that their approach is very valuable to maximise the stability of the learning process, and also for the performance and quality of the learning routes? (p. 241). An even more ambitious attempt [43] to implement intelligence within a PLE through the use of basic AI techniques to personalise the learning environment and provide a much more effective experience. The intelligent PLE automatically managed the content after extracting information from learners interaction, and provided personalised feedback as well as self-evaluation. The authors concluded that the PLE extended the students academic capabilities and boosted cognitive activities. And finally a very recent effort [44] presented a PLE that processes the learners academic history and necessities to extract web resources and customise the content to present a tailored learning environment. Aeiad & Meziane conclude that their approach, functionality and architecture are improvements on existing e-learning systems? (p. 298). The authors state that future versions will attempt to employ details that characterise the individual student as well as their performance to personalise the PLE.

5.4 Conclusion

The personal learning environment or PLE has been analysed in some detail in this chapter together with the two main components, the PLN and PLP. Numerous PLEs have been investigated as different categories and pedagogical approaches have been brought to light to shed numerous insights on the intelligent PLE that will be at the center of discussion in the next chapter. The different PLEs discussed including the intelligent ones are mainly grounded and supported by a foundational institution VLE or LMS upon which additional functionality is appended and incorporated to achieve the required pedagogical effect. The next chapter will introduce and delve into the details of a proper stand-alone intelligent PLE that will build on previous attempts and take on board functional and effective methodologies and concepts that previous PLEs found to be academically valuable and which concur with the methodology and approach being proposed. The intelligent PLE is required to be completely learner-centered with a dynamic and active PLN together with an evolvable and up-to-date PLP. The final intelligent PLE that was mentioned in the previous section suggested that future PLE generation require a feedback cycle to ensure that the personalization process in dynamically in touch with the learner and truly characterise the needs and academic requirements of the same learner. Now that all the required elements and essential components for proposing a truly intelligent PLE an AI-injected e-learning system is presented.

References

1. Morrison, D.: How to Create a Personal Learning Portfolio: Students and Professionals (2013). https://onlinelearninginsights.wordpress.com/2013/01/30/why-students-need-personal-learning-portfolios-more-than-we-do/. Cited 10 March 2015
2. Wheeler, S.: Anatomy of a PLE (2010). http://www.steve-wheeler.co.uk/2010/07/anatomy-of-ple.html. Cited 12 November 2014
3. Leone, S.: Characterisation of a Personal Learning Environment as a Lifelong Learning Tool. Springer, Ancona (2013)
4. Siemens, G.: Connectivism:A Learning Theory for the Digital Age. (2004). http://www.elearnspace.org/Articles/connectivism.htm . Cited 21 April 2014
5. Downes, S.: Places to go: Connectivism and connective knowledge. Innovate. J. Online Educ. 5(1) (2008)
6. O'Reilly, T.: What Is Web 2.0 Design Patterns and Business Models for the Next Generation of Software. (2005). http://www.oreilly.com/pub/a/web2/archive/what-is-web-20.html. Cited 12 Feb 2013

7. Sclater, N.: Web 2.0, Personal Learning Environments, and the future of Learning Management Systems. Colorado: Educause: Center for Applied Research. (2008)
8. Gurzick, D., White, K.: Online personal networks of knowledge workers in computer-supported collaborative learning. In: Goggins, S.P., Jahnke, I., Wulf, V. (eds.) Computer-Supported Collaborative Learning at the Workplace, pp. 225–239. Springer, New York (2013)
9. Baumgartner, P.: Eine Taxonomie fr E-Portfolios, Thesis. Department of Interactive Media and Educational Technologies, Danube University Krems (2012)
10. Gooren-Sieber, S., Henrich, A.: Systems for personalised learning: personal learning environment vs. E-Portfolio?. In: Cheung, S., Fong, J., Kwok, L., Li, K., Kwan, R. (eds.) Hybrid Learning, pp. 294–305. Springer, Germany (2012)
11. Lorenzo, G., Ittelson, J.: An Overview of E-Portfolios. Educase Learning Initiative, Educase (2005)
12. Daunert, A., Price, L.: E-Portfolio: A practical tool for self-directed, refl ective, and collaborative professional learning. In: Harteis, C. e. (ed.) Discourses on Professional Learning: On the Boundary Between Learning and Working, pp. 231–251. Dordrecht, Springer Science+Business Media (2014)
13. Noesgaard, S., Ørngreen, R.: The effectiveness of e-learning: an explorative and integrative review of the definitions, methodologies and factors that promote e-learning effectiveness. Electron. J. eLearning 13(4), 278–290 (2015)
14. Attwell, G.: E-Portfolios the DNA of the personal learning environment? J. eLearning Knowl. Soc. 3(2), 39–61 (2007)
15. D'Alessandro, M.: Connecting your radiology learning to your clinical practice: using personal learning environments, learning portfolios and communities of practice. Pediatr Radiol 41, 245–246 (2011)
16. Yongqiang, H., Jinwu, Y.: Study on the evaluation system of E-Learning based on E-Learning portfolio. ICCIC 2011, pp. 420426. Springer, Heidelberg (2011)
17. Guo, Z., Greer, J.: Electronic portfolios as a means for initializing learner models for adaptive tutorials. EC-TEL 2006, pp. 482487. Springer, Heidelberg (2006)
18. Gauch, S., Speretta, M., Chandramouli, A., Micarelli, A.: User profiles for personalized information access. In: Brusilovsky, P., Kobsa, A., Nejdl, W. (Eds.), The Adaptive Web, pp. 54-89. Springer, Heidelberg (2007)
19. Brusilovsky, P., Milln, E.: User Models for Adaptive Hypermedia and Adaptive Educational Systems. In: Brusilovsky, P., Kobsa, A., Nejdl, W. (Eds.), The Adaptive Web, pp. 3–53. Springer, Heidelberg (2007)
20. Vargas-Vera, M., Lytras, M.: Exploiting semantic web and ontologies for personalised learning services: towards semantic web-enabled learning portals for real learning experiences. Int. J. Knowl. Learn. 4(1), 1–17 (2008)
21. Dagger, D., Wade, V., Conlan, O.T.: Anytime, anywhere learning: The role and realization of dynamic terminal personalization in adaptive learning, Ed-Media 2003: World Conference on Educational Multimedia, Hypermedia and Telecommunications. (2003)
22. Manouselis, N., Sampson, D.: Dynamic knowledge route selection for personalised learning environments using multiple criteria. Intelligence and Technology in Education Applications Workshop - ITEA2002. Innsbruck (2002)
23. Van Harmelen, M.: Personal learning environments. Sixth IEEE International Conference on Advanced Learning Technologies, pp. 815-816. Washington: IEEE Computer Society (2006)
24. Attwell, G.: Personal learning environments - the future of elearning? eLearning Papers 2(1) (2007)
25. Charlier, B., Henri, F., Peraya, D., Gillet, D.: From Personal Environment to Personal Learning Environment, 5th European Conference on Technology Enhanced Learning (EC-TEL 2010). Barcelona (2010)
26. Al-Zoube, M.: E-Learning on the cloud. Int. Arab J. e-Technology. 1(2), 58–64 (2009)
27. Peter, Y., Leroy, S., Leprtre, E.: First steps in the integration of institutional and personal learning environments, EC-TEL (2010)
28. Elgg Foundation: ELGG (2004). https://elgg.org/ Cited 21 Feb 2017

29. Beauvoir, P.: PLEX, Personal Learning Environment (PLE) theme, JISC-CETIS Conference (2005)
30. White, S., Davis, H.: Making it rich and personal: crafting an institutional personal learning environment. Int. J. Virtual Pers. Learn. Environ. **2**(3), 1–18 (2011)
31. Conde, M. A., Garcia-Penalvo, F. J.; Alier, M.: Interoperability scenarios to measure informal learning carried out in PLEs. In: Xhafa, F., Barolli, L., Kppen, M., Proceedings of Third IEEE International Conference on Intelligent Networking and Collaborative Systems, pp. 801–806. IEEE Computer Society. Los Alamitos (2011)
32. Ebner, M., Taraghi, B.: Personal learning environment for higher education - A first prototype, World Conference on Educational Multimedia, Hypermedia and Telecommunications, pp. 1158–1166. AACE. Chesapeake. USA (2010)
33. Garcia-Penalvo, F.J., Conde, M.A., Alier, M., Casany, M.J.: Opening learning management systems to personal learning environments. J. Univers. Comput. Sci. **17**(9), 1222–1240 (2011)
34. Millard, D.E., Davis, H. C., Howard, Y., McSweeney, P., Yorke, C., Solheim, H.: Towards an institutional PLE. PLE Conference (2011)
35. Moccozet, L., Benkacem, O., Platteaux, H., Gillet, D.: An institutional personal learning environment enabler. In: Aedo, I., Bottino, R., Chen, N., Giovannella, C., Sampson, D., Kinshuk. In: Proceedings of the 2012 12th IEEE International Conference on Advanced Learning Technologies, pp. 51–52. Rome: IEEE Computer Society (2012)
36. Casquero, O., Portillo, J., Ovelar, R., Benito, M., Romo, J.: iPLE network: An integrated elearning 2.0 architecture from a universitys perspective. Interact. Learn. Environ. **18**(3), 293–308 (2010)
37. Dabbagh, N., Kitsantas, A.: Personal learning environments, social media, and self-regulated learning. Internet High. Education **15**, 3–8 (2012)
38. Fiedler, S.H.D., Vljataga, T.: Personal learning environments: a conceptual landscape revisited, eLearning papers n. 35 (2013)
39. Valtonen, T., Hacklin, S., Dillon, P., Vesisenaho, M., Kukkonen, J., Hietanen, A.: Perspectives on personal learning environments held by vocational students. Comput. Education. **58**, 732–739 (2012)
40. Pearson, E., Gkatzidou, V., Green, S.: A proposal for an adaptable personal learning environment to support learners needs and preferences, ascilite Auckland 2009, pp.749–757. Australia (2009)
41. Cui, X., Zhang, S.: The Personalized E-Learning System Based on Portal Technology in Zhiguo, CCIS, pp. 433–439. Springer, Heidelberg (2011)
42. Morales, L., Garrido, A., Serina, I.: Planning and execution in a personlaised e-learning setting, Lecture Notes in AI, pp. 233–242. Springer, Berlin (2011)
43. Xu, D., Huang, W.W., Wang, H., Heales, J.: Enhancing e-learning effectiveness using an intelligent agent-supported personalised virtual learning environment: An empirical study, Information and Management. pp. 430–440 (2014)
44. Aeiad, E., Meziane, F.: An adaptable and personalised e-learning system based on free web resources in Biemann, NLDB 2015, LNCS 9103, pp. 293–299. Springer, Switzerland (2015)

Chapter 6
Customised e-Learning – A Proposed Model

We keep moving forward, opening new doors, and
doing new things, because we're curious and
curiosity keeps leading us down new paths.

Walt Disney

Abstract This chapter brings all the previous chapters together as they collectively and incrementally built up a crescendo to reach the highlight, namely injecting e-learning with AI to customise the education process. The proposed model makes us of all the techniques discussed in the previous chapters and endeavours to compatibly bring them together to create an intelligent personal learning environment. The evolution of e-learning was led imposed by the technology but this model proposes to conveniently employ numerous technologies and techniques to directly address specific e-learning issues. The next generation of online education is not dictated by technology but by the academic need to personalise learning together with the efficient automation offered by AI. The first e-learning issue addressed is that of isolation and Chap. 3 undertook this task with the ingenuity of crowdsourcing and the popularity of social networks. The connectivism learning theory has been associated with this phenomenon and this model makes good use of this first factor. Motivation is another e-learning issue that is addressed through the contributions from Chap. 4 as learner profiling and learning portfolios support student to be much more self-determined in their academic endeavour. The third and final issue tackles the issue of impersonality that e-learning is notoriously criticised, and Chap. 5 offers adaptive environments through the combination of a learning portfolio and supportive learning network. A truly intelligent personal learning environment backed and injected by AI techniques is being proposed as a compatible combination of all these technologies to enhance e-learning effectiveness as it leads online education to its future and the next e-learning generation. The rest of this chapter is organised as follows. The section that follows expands further the underlying rationale that led to the proposed model by analysing the contributions from the previous chapters. This is followed by the architectural setup of how these technologies come together within an online system to deliver a functional and intelligent PLE. The next section tackles all the implementation details that take place to accomplish and complete the architectural design presented before. Finally operational and pragmatic details of

51
M. Montebello, *AI Injected e-Learning*, Studies in Computational Intelligence 745,
https://doi.org/10.1007/978-3-319-67928-0_6

how the online PLE functions are covered in an effort to show how the AI-injected e-learning system will operate in reality.

6.1 Background Rationale

The rationale behind the proposed model is grounded on the hypothesis that through the use of AI techniques the next generation of e-learning platforms are far more effective as personalisation, learner profiling, and social networks come together in a functional personal learning environment. These techniques are based on respective learning theories in an attempt to address numerous e-learning concerns that currently limit the potential success and deserved accomplishment of online education. The model is particularly intricate due to its multifaceted nature as it adopts and applies a number of techniques in a well-thought and defined combination. The established methods resulted from a bottom-up analysis of current e-learning practices as discussed in Chap. 2, as a result of which counteractive measures together with sound justifications are able to address such issues and even more promise to add value and enhance online education. A learning environment through its different components as discussed in the previous chapter lends itself perfectly to create a healthy personal academic eco system that can address the e-learning concerns that have unfortunately held back the full potential of online education. The next three subsections re-present the PLE and its sub components from an alternate point of view as they home solutions to corresponding issues and intelligent techniques supported by respective learning theories. The combination of these elements are at the basis of this model and focus mainly on the application of AI techniques to achieve a level of sophistication as close as possible as what one expects from a personal tutor. The fact that known e-learning issues are being addressed within this proposal while at the same time established learning practices are compatibly merged together to offer a solution within an academically favourable environment renders the model valid and achievable.

6.1.1 Incorporating a Social Aspect

The first pillar of this solution proposes a social aspect which integrates perfectly with the learning network discussed in Sect. 5.1. As highlighted earlier Leone [1] highlights the positive influence of support system around the learner as it offers multiple sources of heterogeneous information while presenting alternate points of view. Associated with this first component are three inter-related concepts that propound a complete and fitting resolution, namely reducing isolation issues, employing social media and crowdsourcing techniques, and the learning theory of connectivism.

Introducing a social aspect within an e-learning environment is the closest one can get to the communal atmosphere within a physical classroom. Issues of isolation reported during e-learning [2–4] are challenged as the learner develops a healthy and mutual working relationship with fellow learners and web users. A personal learning network empowers each of the learners to not only build a self-esteem of initiating, building and cultivating such an academic network, but also the ability and sense of achievement of sharing, collaborating and sourcing new information, ideas and knowledge. The commitment to do so is no simple task or a one-time impulse to register and become a member of a network or learners, but an ingrained mind-set and a conscious way of life. It is this belonging and dedication that fosters a communal sense of belonging and non-isolation. The massive knowledge-base found on the WWW together with all contributing users, domain experts and educators offer an incredible potential at the fingertips of every learner which automatically imparts an intense and overwhelming sense of control, cooperation and collaboration. Every learner has a unique approach when developing and enriching ones PLN with respective personal reasons, pace and motivations. In this way a healthy educational eco-system flourishes and strengthens as like-minded learners assist each other as they recursively collect, distribute, curate and generate further content in a collaborative and cooperative online environment. Adequate software tools are required to initiate and curate a personal learning network as networking skills and information harvesting abilities are a necessity for a learner to contribute to other learners as much as or even more than acquiring from them who likewise are cultivating their own PLN. It is this community sense and group atmosphere that the model aims to achieve in an effort to reduce situations where learners find themselves isolated within the e-learning environment and feeling alone without the comfort of others around.

Web 2.0 has also contributed to the emergence of numerous networking tools [5, 6] that dynamically foster the proliferation of this eco-system as learner-generated content, activities and initiatives are encouraged and advocated. The use and engagement of online sources that the general public provides is also referred to as crowdsourcing as mentioned in Chap. 3, and its employment in higher education forms part of the proposed model as it mines and taps into the massive knowledge-base of online content that is open and relevant to the different academic needs of different learners. The use of crowdsourcing is employed similar to the way a number of research projects made good use of online resources that include academic content, tutors, blogs, domain experts and social networks in an effort to add value and enhance the effectiveness of the e-learning system. Costa et al. [7] reported a rise in active learning during a project that employed a knowledge-base to source their content and which was populated through crowdsourcing. Their conclusions distinguished between generic content that performed better than standard and traditional educational techniques and specifically focussed content which would require some kind of human intervention or intelligence to tailor the material to the specific needs of the learners. If the academic content is purposely tweaked to the special needs of a learner then the crowdsourcing process turned out to be much more accurate and effective. This implies that the issue here is not the crowdsourcing process itself that has an issue but the personalisation of the content, and thereby it follows that crowdsourcing on its own is not enough and needs additional techniques to accompany it as proposed in the model. Weld et al. [8] arrived at the same conclusions and reiterated that such a combination could unleash the true potential of e-learning. Another reason for employing crowdsourcing as part of the model is the unbiased nature of the content emanating from different and diverse sources. Such a diversity is encouraged by some researchers [9] and does not mean that the content is not organised, as structured and indexed educational knowledge-bases like Open Educational Resources (OER) [10] and Merlot [11] provide interfaces to efficiently access online content. Other online Web 2.0 tools [12] facilitate the process and make it possible to access aggregated content that has been crowdsourced and indexed ready to be used within a learning environment.

The third concept upon which the PLN component is based upon centres around the connectivism learning theory that corroborates the claims of alleviating isolation issues, and substantiates the use of crowdsourcing. As mentioned earlier in Chap. 4 this learning theory has been associated by several researchers [13–15] to the digital age as it takes into consideration online learners as they connect through their learning networks. Others [16] take it a step further and refer to this theory as a crucial theory that justifies how learners personal lives are dependent on information which is sourced from their online connections. An academic study [17] reported on how the connectivism learning theory was employed to enhance a learning process through the integration of Web 2.0 applications. Researchers [18] attribute this theory to the changing nature of e-learning from the underlying cognitive, instructivist and behaviourist pedagogical learning theories to a social, constructivist and connectivist. Other researchers [15, 19] take advantage of this theory to investigate diversity in learning through different networks and in conjunction with other learning theories. In this way new models of learning are designed to investigate how connected knowledge can be managed and transferred, and how learning spaces and structures can change and expand in an attempt to connect learners through open technologies. For these reasons this model does not engage this learning theory in isolation but in conjunction with others as presented in the next two subsections, simply because the learning environment is complex and multi-faceted, and other issues need to be taken into consideration. In agreement with this argument numerous researchers [15, 19, 20] point out that teaching contexts like asynchronous sessions need to be justified by more than the connectivism learning theory. Others [21] insist that learning in this digital age is nothing special and no one learning theory could possibly applied to cover all the different online learning methodologies employed. The proposed model however makes extensive use of social networks and crowdsourcing and thereby the connectivism learning theory is the principal learning theory.

These arguments justify the amalgamation of all these concepts, addressing isolation while adopting social networks through crowdsourcing and justified by the connectivism learning theory, within the personal learning network component. These concepts compatibly blend together supporting each other as they jointly offer fruitful learning outcomes. If one had to compare this PLN component to a student-teacher relationship whereby a true educator would revert to any possible teaching methodology, aid, and medium to optimise the learning channel. Such an ideal scenario is not enough in isolation as it still requires a functional and efficient communication chan-

nel without any barrier whatsoever. In a parallel dimension crowdsourcing, social
networks and online content are not enough and would not suffice without the right
setting, compatible networkers and the correct attitude towards the whole process.
Over and above this learners within each others PLNs and consuming content need
to provide content as they recursively contribute to this cycle by their comments,
blogs, posts, tweets, discussions, pictures and all other content sourced from Web
2.0 applications. The PLNs main concern is to optimise the content crowdsourcing
mechanism to populate the learners knowledge-base from freely available content
providers and domain experts that would otherwise be hard to collect. The final outcome that the PLN component provides is a complete online resource center that
incorporates content together with connections to other content and knowledge.

6.1.2 Augmenting Learner Motivation and Determination

The learners motivational levels are addressed in the second pillar as the personal
learning portfolio component is employed to maintain an updated profile and a constant determination to learn and participate within the learning environment. The PLP
component also has three associated inter-related concepts that will be expanded further to justify their essential contribution to the proposed model. The concepts that
factor in and feature within the PLP component are the learner profiling practices
that assist in raising the learners motivation and which are supported by the self-
determination learning theory.

Learner profiling has been expanded in Chap. 4 and referred to in Sect. 5.2 as part
of the PLP component due to its central bearing within the proposed model. Such
a profile will enable the customisation of the content that has been made available
by the PLN from the previous subsection. The model employs learner profiling
to ensure that the customisation process of all the content aggregated by the learner
network is duly tailored and relevant to the specific learner. Once the learner interacts
with the model a feedback loop will ensure that the PLP is further refined and fine-
tuned to fit even better the learners academic needs and requirements. The model
makes use of different established machine learning techniques that provide the
required learner profile and which can easily be replaced by better and efficient future
profilers. Technical details of the profiler are not the novelty here but the setup of
the model bringing together numerous concepts, ideas and techniques in a functional
and effective personal learning environment. Different profiling techniques function
differently with ultimately the same aim, and their performance varies from one
to another depending on the contributing factors as well as the internal setup of
the machine learning technique. The essential thing here is that learner profiling
is provided through the processing of learner interaction with the environment that
generates data and input to the same profiler that is eventually employed to customise
the content generated by the PLN. In conjunction with the learner profile generated
the model makes use of clustering techniques that are also artificially intelligent
processes that match the learner profile with other profiles that could potential be

similar or closely correlated. The proposed model associates the unique learner profile with one or more clusters in an effort to propose and recommend customisation of the learning environment based on different aspects of the learners distinct profile.

A learner who experiences an environment that responds to the academic needs, interests and requirements becomes even more motivated to pursue and actively feel part of the educational process. Such a lack of determination that results from a drop in enthusiasm has been at the centre of a study [22] where the learner response and feedback resulted to be critical and central in the choice of adequate course content and which eventually helped in raising motivational levels. Deci and Ryan [23] associate motivation with self-determination and have purposely come up with a learning theory to address the learners demeanor in an effort to improve their academic performance. The self-determination learning theory specifies a variety of motives and objectives that activate different kinds of learner motivation. A learner could be intrinsically motivated if the action or experience ensues from an action or situation that in itself was intrinsically enjoyable, pleasing and appealing. It could also be the case that a learner is extrinsically motivated as the learning situation leads to a distinct and exclusively enjoyable outcome. Motivational levels and self-determination are directly linked to the process of learner profiling as specific learner characteristics and learning patterns are identified and employed to customise the academic content to satisfy the same learner. Researchers [24, 25] have shown that personality traits are correlated to academic performance which the PLP ultimately encapsulates. The profile generated and represented within the learning portfolio is characteristic of the specific academic profile of a student that eventually functions as a motivational channel to enhance the learning experience. Another study [26] highlights the fact that the current interests and needs of a learner have a central role and are instrumental in the learners motivational levels and the corresponding academic success. This reinforces the reasoning behind the dynamic PLP that is continuously refined through the feedback cycle driven by the learners interaction with the tailored content. These studies support and justify the rationale behind this second component of the model whereby motivational issues are addressed through the learner profile generation and application as advocated by the self-determination learning theory with the PLP. Additionally other researchers [27, 28] associate also self-regulated learning to self-determined and self-directed concepts together with non-linear learning. The learner is motivated to understand and appreciate how to learn as control over content, environment and process empower and induce the learner to perform even better.

6.1.3   Personalising the Learning Environment

The previous two components come together to form a learning environment that incorporates the techniques employed in the PLN and PLP as well as personalises the entire learning experience into a PLE. The personal learning environment discussed in Sect. 5.3.3 is at the bases of the proposed model as AI is injected into the learning environment through the techniques employed in the other two constituent parts. This customised interface provides the much required personalization that addresses another e-learning issues of the learning environment being impersonal, and which subscribes to another learning theory, the adaptive learning theory. These three inter-related concepts will now be expanded in the context of the PLE.

The personalization concept as discussed in Chap. 4 is to be commended as its beneficial effects within an e-learning environment are purposely incorporated in the model to add value to the overall effectiveness and academic significance. The potential of personalised learning in relation to available online content was previously explored [29] with the help of networks to aggregate the information over the web in a rational and significant way. The proposed model employs personalization in more than one aspect that other e-learning attempts, mentioned earlier in Sect. 4.1, employed in isolation. The physical environment of the proposed model offers a degree of personalization to help the learner associate oneself with the surroundings. Such aesthetics reflect the learners self-expression by the way items, content, icons and all visual aspects appear within the interface. Another aspect of how the model offers learner personalization is through the different content providers selected, as well as the particular services and functionalities the learner elects to integrate within the same learning environment. Basically the PLN can be tweaked and attuned to each specific learner and according to accepted recommendations that will contribute to the compilation of the PLP. Finally the model allows the possibility of supplementing all the implicit information gathered by the PLP with explicitly indicated interests, needs and requirements which can be catered to immediately as the PLP continuously updates and refines its filtering mechanism. All three aspects of personalization are not AI-based but still offer a customised representation of the model to every unique learner, which is what the PLE is meant to deliver.

The personalization process of the PLE addresses the third and final e-learning issue, that of an impersonal e-learning environment. Online learners continuously complain that the environment is cold and void of any human interaction thereby creating an ineffective academic environment. The proposed model makes use not only of a PLE but also of an intelligent element fueled by AI techniques and capabilities. To such ends the intelligent PLE that incorporates both the personal learning network and portfolio creates a customised and friendly learning environment that is appealing to the user and which renders the e-learning process even more effective.

A learning theory that endorses the personalization of the learning environment is the adaptive learning environment as it supports the value-added notion of customizing and tailoring learning content and methodologies to the needs, interests, and requirements of the learner. The educational added value that personalizing the learning process generates is no secret or has never been attempted [30], but the way to achieve and maintain such a process is not straightforward. In the last ten years adaptive learning system have also been developed by others [31–36] who have reported the potential and pervasiveness of these effective systems at all levels of education. The personalization process is required to be complete and comprehensive and not just aesthetic, thereby including customization of the content, methodology and processes [37]. Similarly other educational researchers [38, 39] adopt the same ideology and argue that the learners uniqueness needs to be coupled with the contextual relativeness in time of the same learner as interests, needs and requirements continuously change and evolve. Three factors emerge from all these studies that together constitute an integral part of the proposed model. The learners interests and academic needs are dynamically changing both through a maturing learner but also through the evolution of the knowledge processes that occur through education. Secondly the learner has a combination of interests and thereby the model caters for one or more domain area rather than a single monolithic model. Finally, the content and education material is also a combination of topical interests with more than one dimension and facet. The proposed model takes all these factors into consideration in order to ensure a truly personalised and contextualised learning environment. It adopts a triangular structure whereby the content, the learner and the instructional methodology contribute to the academic parallel in direct correspondence with the PLN, PLP and PLE respectively.

6.2 Architectural Design

The proposed model brings together numerous concepts apart from various technologies required to implement the methodologies researched, presented and discussed. Each feature of the proposed model as introduced in the previous subsections are depicted in Fig. 6.1 reflecting all three aspects of the intelligent PLE that embodies the AI-injected e-learning platform to personalise online learning as we know it. The three learning theories are at the core of the model as they embody the philosophical reasoning behind the architectural setup. The learning theories compatibly come together and as justified earlier it is common practice within this domain to coalesce theories supporting each other and converge together towards one goal. They are somewhat related to each other as the focus is taken off the educator and predominantly focuses on engaging learners and their peers. The three learning theories are intentional aimed to address the three e-learning concerns shown here at the vertices of the personal learning concepts triad. As argued earlier the proposed model addresses these concerns at the root of classical e-learning systems while proposing an architectural rearrangement that could define the future of e-learning. The integration of the three techniques, Crowdsourcing, Learner Profiling, and Personalisation, clearly featured in underlying rationale figure, form the inner triangular structure upon which the proposed model is assembled in a single e-learning platform. The injection of AI permeates through all these techniques to address each concern and offer a functional and effective solution. The proposed e-learning environment is therefore intended to address the following three matters:

- Ensure that no learner is left isolated, struggling to cope alone and muddle through a standard one-size-fits-all course programme. Online learners can form part of a learning network in an effort to integrate within a comprehensive learning society whose members are sources of information as much as they are recipients. This

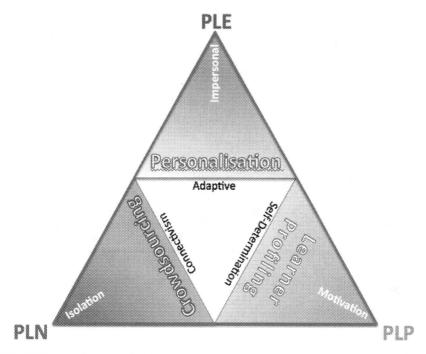

Fig. 6.1 Proposed model underlying rationale

matter is deeply rooted within the connectivism learning theory that focuses on a digital society where every individual learner is not abandoned in isolation but forms part of a healthy network of academic nodes. The personal learning network or PLN that each learner develops, curates and possesses embodies this effort and forms part of the proposed environment. Use of clustering AI techniques further augment in the effectiveness of such PLN formations as like-minded learners are automatically brought together as they can mutually benefit from each others learning experience.

• Assist in continuously maintain the learners motivational levels as high as possible throughout the learning process. This is achieved through the correct recognition of what the learner is interested in and by ensuring that the specific learning processes are accurately captured in an efficient unobtrusive way that represent patterns in the needs, interests and enthrallments of the same learner. The self-determination learning theory addresses these concerns as motivated learners are empowered to completely control their own learning processes, and the personal learning portfolio or PLP component encapsulates this effort. Learner profiling AI techniques are employed to specifically cater for this particular matter.

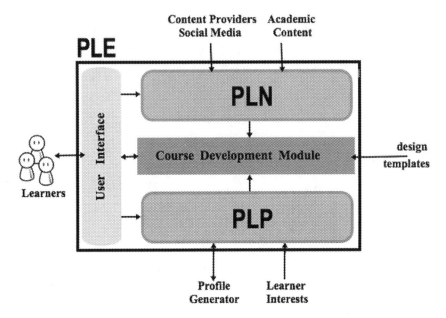

Fig. 6.2 Top-level architecture

- Finally both matters come together to conjointly contribute in the personalization of the learning process and the entire e-learning experience. This is possible
  through the proper and accurate tailoring of the academic content and processes
  while ensuring that the customised medium optimises the delivery of the online
  programme. The adaptive learning theory specifically addresses this issue by ensuring that the learning process is not impersonal or detached from the learner but
  meticulously and automatically adapted to each individual learner. The personal
  learning environment or PLE incorporates the previous two components, PLN and
  PLP, epitomises the overall model being proposed while probing into the future
  of online education and the e-learning generation.

The above three objectives delineate the architectural design of the proposed
model as each one of them accurately demarcates all that has been stated in the
previous chapters in a build-up towards the proposed model. The top-level architectural design shown in Fig. 6.2 depicts how the different components come together
to make up an intelligent PLE.

6.2.1 Intelligent User Interface

The learner interacts with the system through an embellished intelligent user interface that is customised and purposely dynamically generated in real-time to truly personalise the e-learning experience, aesthetically, content-wise, and functionally. The user interface is intended to optimise the learners interaction throughout the e-learning sessions and programme. The look-and-feel is easily controlled by the use of design templates to consistently maintain what the learner prefers and feels comfortable with. This also gives a sense of modularity and ease to evolve as new and diverse interfaces are imported to satisfy the tastes of different learners.

6.2.2 Real-Time Course Development

At the centre of the intelligent PLE design is a course development module that generates the academic course itself as it takes input from both the PLN and PLP. The course content is personalised accordingly and formatted according to the design template that fits the particular learner. The PLN, as explained earlier, is seen here within the PLE as it provides the required input to the PLE while it provides its own content from three components, crowdsourcing, a pre-set academic programme, and feedback from the learners interaction with the learning environment. All three inputs are dynamic in nature and modularly interchangeable as the learner can edit, add and remove new sources, functionality, widgets and content at will. Similarly the PLP contributes to the generation of the personalised e-learning course development while taking its own input from three components. Learner interactions and feedback loops are the most important input to the PLP that engages a profile generator to renew and update the current learner profile, while keeping track, adjusting and refining the constantly changing profile encapsulated within the PLP. Finally the learners interests need to follow and closely represent the true and real current interests of the learner to validly contribute to the PLP being continuously generated and updated in real time.

6.2.3 iPLE

The intelligent PLE or iPLE is intended to automatically suggest, advice and recommend new sources, functionalities and related interests that become available without imposing or obliging the learner to adopt or accept. However should the learner decide to accept any of the suggested recommendations, they will be included and reflected within the dynamic course development process. The intelligent PLE is meant to reversibly accommodate the learners needs, interests and requirements while concurrently presenting the academic content pitched at the correct academic level of the specific learner.

6.3 Development Aspects

The development of the proposed model goes beyond the scope of this book as several technologies, development environments and implementation possibilities are available. In this respect it is worth mentioning specific aspects that need to be taken into consideration irrespective of which technology is employed to develop the intelligent PLE. The proposed learning environment is online and assumes online access to WWW resources. The best option, typical to majority of e-learning systems, is to develop a web-based application and thereby employ a web browser as the user interface. Apart from the ease to access the learning environment from any location or any machine, due to the different versions of web browsers over diverse platforms, numerous other issues make the adoption of this option most advantageous. No downloading whatsoever is required and no installation issues or hard-disk space need worry the learner as access through the web browser eliminates all these issues. This brings on the added advantage of not needing to update the software accept for the web browser itself. Any new versions or maintenance updates are done on the server-side and thereby not a concern to the learner. This includes any new services, novel capabilities, additional Web 2.0 functionalities, latest online content sources, and up-to-date open educational resources. Such a thin client methodology with a web browser as the interface places the focus on the server performance, the efficiency of the web programming language employed to develop the intelligent PLE, and the performance of the underlying procedures that include the profile generation and updating, the structuring of the content according to the learners PLP, and the actual dynamic rendering of the e-learning session itself.

The real-time course development as specified in Sect. 6.2.2, is the component that requires most attention and that will consume much of the processing time. To alleviate this task it is important to ensure that all the contributing components are either pre-processed or continuously execute in the background in order to have all their required input ready at hand. The design templates and the academic content are probably the easiest of all the inputs as they can easily be made available beforehand with the exception that the adequate content is required to be identified to fit the learners PLP. The templates identify the different positioning of all the components that will be made available including content from crowdsourcing sources related to the content identified. This does not exclude the capabilities of drag-and-drop to rearrange items on the interface apart from additional functionality available that can be added in at will. The latest PLP generated by the profile generator will be employed and thereby no processing time needs to be consumed in such a process. Important to point out that any interactions and feedback given by the user needs to processed concurrently in real-time on the server-side by the same profile generator in order to generate the latest version of the PLP. This process should be scheduled not only every time the learner end an e-learning session, but also every time the learner completes an academic section and is ready to move to the next. Once an academic module needs to be dynamically developed the latest version of the updated PLP is employed together with the most recent declared or selected learner interests to extract content from the diverse providers and social media to adapt and customise the targeted academic content. To clarify that the academic content is the only matter that an educational institution or the academic administrator pre-declares and specifies in well defined categories to facilitate the work of the course development module.

The learners feedback, progress, performance, and choices are duly recorded on the server-side along with the credentials and PLE details of the same learner. The intelligent PLE can easily employ a registration methodology with client-side cookies to assist the learner of the proper username and password, but the rest of the recording that needs to be done is performed on the server-side. The learner has access to numerous sources that are provided within the intelligent PLE by the crowdsourced content and this enables a mature audience to venture out of the learning environment to other online locations that are closely related to the content and the individual learner interests. The concept behind the intelligent PLE is not just that it is learner-centric, but also that the learners are entirely responsible for their education and commitment towards pursuing their own education, and thereby it is most suitable to a mature target audience from a higher education spectrum.

6.4 Conclusion

The proposed model of customised e-learning was presented in the form of an intelligent personal learning environment or iPLE following a detailed build-up in the previous chapters. The iPLE is made up of different components, technological concepts, pedagogical ideas, and epistemological standpoints that have been clearly discussed and justified. The assembly and organization of the model exposes the compatibility of all these different elements that factor in towards the ultimate aim of personalizing e-learning. The use and integration of AI in every aspect of the model is crucial to this same aim as human intervention to assist, design, tailor and successfully deliver effective e-learning to every individual learner is practically impossible and unrealistic. Artificial intelligence techniques are able to capture the learners individual characteristics and use them to personalise the e-learning delivery while addressing typical e-learning issues of isolation, motivation and impersonal environment. These concerns have not only been addressed through the intervention of different methodologies like crowdsourcing, learner profiling, and personalization, but have been embodied within the PLN, PLP and PLE notions, and thoroughly justified and sustained by respective learning theories of connectivism, self-determination, and adaptivity.

References

1. Leone, S.: Characterisation of a Personal Learning Environment as a Lifelong Learning Tool. Springer, Ancona (2013)
2. Noesgaard, S., Ørngreen, R.: The effectiveness of e-learning: an explorative and integrative review of the definitions, methodologies and factors that promote e-learning effectiveness. Electron. J. eLearning **13**(4), 278–290 (2015)
3. ODonoghue, J., Singh, G., Green, C.: A comparison of the advantages and disadvantages of IT based education and the implications upon students. Interact. Educ. Multimed. **9**, 63–76 (2004)
4. Olson, J., Codde, J., deMaagd, K., Tarkleson, E., Sinclair, J., Yook, S.: An Analysis of e-Learning Impacts and Best Practices in Developing Countries. Michigan State University, Michigan, USA (2011)
5. O'Reilly, T.: What Is Web 2.0 Design Patterns and Business Models for the Next Generation of Software (2005). http://www.oreilly.com/pub/a/web2/archive/what-is-web-20.html. Cited 12 Feb 2013
6. Sclater, N.: Web 2.0, Personal Learning Environments, and the future of Learning Management Systems. Educause: Center for Applied Research, Colorado (2008)
7. Costa, J., Silva, C., Antunes, M., Ribeiro, B.: On using crowdsourcing and active learning to improve classification performance. In: 11th International Conference on Intelligent Systems Design and Applications (ISDA), pp. 469–474. IEEE, Cordoba (2011)
8. Weld, D.S. et al.: Personalised online education - a crowdsourcing challenge. In: AAAI Workshops at Twenty-Sixth AAAI Conference on AI. [S.l.]: AAAI, pp. 159–163 (2012)
9. Bonabeau, E.: Decisions 2.0: the power of collective intelligence. MIT Sloan Management Review, pp. 45–52 (2009)
10. Open Educational Resources (2017). https://www.oercommons.org/. Cited 13 June 2016
11. Merlot II: Multimedia Educational resource for Learning and Online Teaching (2017). https://www.merlot.org/merlot/index.htm. Cited 24 Nov 2016
12. ConsiderIt (2016). https://consider.it/. Cited 11 Nov 2016
13. Downes, S.: Places to go: connectivism and connective knowledge. Innov. J. Online Educ. **5**(1) (2008)
14. Kop, R., Hill, A.: Connectivism: learning theory of the future or vestige of the past? Int. Rev. Res. Open Distance Learn. **9**(3) (2008)
15. Duke, B., Harper, G., Johnston, M.: Connectivism as a Digital Age Learning Theory. The International HETL Review (2013)
16. Urea, G.V., Valenzuela-Gonzalez, J.R.: Online social network contacts as information repositories. Int. J. Educ. Technol. High. Educ. **8**(1), 142–155 (2011)
17. Loureiro, A., Bettencourt, T.: Immersive environments a connectivist approach. In: Lytras, M. (ed.) WSKS 2010, Part I, CCIS 111, pp. 202–214. Springer, Berlin (2010)
18. Robson, R.: The changing nature of e-learning content. In: Huang, R., Spector, J. (eds.) Reshaping Learning - Frontiers of Learning Technology in a Global Context, pp. 177–196. Springer, Berlin (2013)
19. Hung, N.M.: Using ideas from connectivism for designing new learning models in vietnam. Int. J. Inf. Educ. Technol. **4**(1), 76–82 (2014)
20. Ng, W.: Theories underpinning learning with digital technologies. In: Ng, W. (ed.) New Digital Technology in Education, pp. 73–94. Springer International Publishing, Switzerland (2015)
21. Mayes, T., de Freitas, S.: Technology enhanced learning: the role of theory. In: Beetham, H., Sharpe, R. (eds.) Rethinking Pedagogy for a Digital Age: Designing for 21st Century Learning, p. 1730. Routledge, New York (2013)
22. Tang, T., McCalla, G.: Beyond learners interest: personalized paper recommendation based on their pedagogical features for an e-learning system. Lect. Notes Comput. Sci. **3157**, 301–310 (2004)
23. Deci, E., Ryan, R.: Intrinsic Motivation and Self-Determination in Human Behavior. Plenum Press, New York (1985)

24. Chue, K.L.: Examining the influence of the big five personality traits on the relationship between autonomy, motivation and academic achievement in the twenty-first-century learner. In: Koh, C. (ed.) Motivation, Leadership and Curriculum Design, pp. 37–52. Springer, Singapore (2015)

25. Deci, E.L., Vallerand, R.J., Pelletier, L.G., Ryan, R.M.: Motivation and education: the self-determination perspective. Educ. Psychol. **26**(3, 4), 325–346 (1991)

26. Linnenbrink, E.A., Pintrich, P.R.: Motivation as an enabler for academic success. Sch. Psychol. Rev. **31**(3), 2002 (2002)

27. Wheeler, S.: Theories for the digital age: Self regulated learning, Learning with 'e's (2012). http://www.steve-wheeler.co.uk/2012/10/theories-for-digital-age-self-regulated.html. Cited 12 Nov 2014

28. Hase, S., Kenyon, C.: Heutagogy: a child of complexity theory. Int. J. Complex. Educ. **4**(1), 111–118 (2007)

29. Siemens, G.: SenseMaking Artefacts, Connectivism (2012). www.connectivism.ca. Cited 15 Apr 2016

30. Oxman, S., Wong, W.: Adaptive Learning Systems. DeVry Education Group, USA (2014)

31. Soonthornphisaj, N., Rojsattarat, E., Yim-ngam, S.: Smart E-learning using recommender system. Comput. Intell. **4114**, 518–523 (2006)

32. Drachsler, H., Hummel, H., Koper, R.: Personal recommender systems for learners in lifelong learning: requirements, techniques and model. Int. J. Learn. Technol. **3**(4), 404423 (2008)

33. Kay, J.: Lifelong learner modeling for lifelong personalized pervasive learning. IEEE Trans. Learn. Technol. **1**(4), 215–228 (2008)

34. Tan, H., Guo, J., Li, Y.: E-learning recommendation system. In: International Conference on Computer Science and Software Engineering, pp. 430–433 (2008)

35. Bobadilla, J., Serradilla, F., Hernando, A.: Collaborative filtering adapted to recommender systems of e-learning. Knowl.-Based Syst. **22**(4), 261265 (2009)

36. Bian, L., Xie, Y.: Research on mutual adaptation problem of adaptive learning systems. J. China Educ. Technol. **3**, 9–12 (2009)

37. Bian, L., Xie, Y.: Research on the Adaptive Strategy of Adaptive Learning System. In: Zhang, X.E. (ed.) Edutainment, pp. 203–214. Springer, Heidelberg (2010)

38. Oxman, S., Wong, W.: Adaptive Learning Systems. DeVry Education Group, USA (2014)

39. Salehi, M., Kamalabadi, I.N., Ghaznavi Ghoushchi, M.B: Personalized recommendation of learning material using sequential pattern mining and attribute based collaborative filtering. Educ. Inf. Technol. **19**, 713–735 (2014)

Chapter 7
Looking Ahead

We cannot solve our problems
with the same thinking
we used when we created them.

Albert Einstein

Abstract The book draws to an end by looking ahead at potential future avenues in light of the proposed intelligent personal learning environment. Web technologies and AI techniques continue to evolve as e-learning systems continue to take full advantage of both to improve the delivery and the overall holistic experience. The employment of AI techniques in combination with other technologies moved away from the conventional trend of adopting the latest web technologies to embellish the e-learning environment and move to the next generation. The proposed intelligent learning environment had set objectives with specific issues to resolve and employed the different methodologies and practices within an original architectural setup that fulfils the pre-set e-learning needs. Will it be possible to pursue this trend whereby the e-learning needs dictate and prescribe what the technology should be like and impose what it should provide? On the other hand the same architectural setup introduced a novel concept of bringing together numerous technologies to achieve a common goal, personalised e-learning. Will future e-learning generations persist on this line of thought and take full advantage of multiple developments in numerous and diverse domains to collectively achieve a superior added-value outcome that could potentially shape the future of e-learning? This final chapter looks ahead at these possibilities and the potential of influencing future e-learning generations by reversing the way e-learning advocates reason and devise such futuristic environments. Which technological novelties will leave their impact on future e-learning setups? What exactly is the ideal e-learning scenario and which technologies or combination of technologies can pave the way forward?

© Springer International Publishing AG 2018   67
M. Montebello, *AI Injected e-Learning*, Studies in Computational Intelligence 745,
https://doi.org/10.1007/978-3-319-67928-0_7

7.1   Web 3.0

The proposed model that makes extensive use of AI in combination with other techniques to address numerous e-learning concerns and substantiated by respective learning theories realises the next step in e-learning evolution as numerous researchers speculated about in the last few years. Rubens et al. [1] also identify AI as a driving force and hypothesise that machine learning and data mining techniques are potential technologies to enable a shift towards Web 3.0 and the next version of e-learning. The authors predicted that AI would exert a major influence on the development of e-Learning 3.0 as it commonly referred also by other e-learning advocates [2] who also claim that Web 3.0 will be the catalyst to the next generation e-learning (See Fig. 7.1). The AI represented in the form of intelligent agents and semantic standards is argued to be crucial in revolutionizing the academic approach online together with virtual spaces and 3-D visualizations. The same researchers claim that a major transformation in e-learning systems is required in order to enhance online interactions between learners and potential educational resource while eradicating the passive methodology of accepting and remembering explicit information passed on by others. Hussain [3] not only concurs but establishes a relationship between the third generations of e-learning and web as she argues that e-Learning 3.0 is an extension or a simple transformation of the previous e-learning generation with the added technological input of Web 3.0 but fails to make any reference to any AI technologies that might contribute. This trend, highlighted in Chap. 2, persists with other researchers [4] who theoretically pair up respective generations of web and e-learning while accrediting the evolution of the latter due to the advancements of the prior. Wheeler [5] identifies four technologies that he considers key drivers to e-learning 3.0, namely, are distributed computing, extended smart mobile technology,

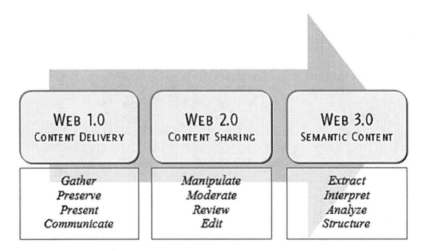

Fig. 7.1 Web Evolutions [2]

collaborative intelligent filtering, and 3-D visualization. How can these technologies come together to deliver on their promise? But why should the next e-learning generation be at the mercy of technological development? Future generations of e-learning should prod ahead independently of any technological restrictions or advancements and unreservedly and liberally evolve based on strong and sound learning theories compelling the required technological support to accommodate the new e-learning developments.

7.2   Ambient Intelligent Learning Environment

In an attempt to envisage the next e-learning generation without being influenced by any technological developments I will attempt to build on the AI-injected e-learning model presented while moving away from the web-based that unfortunately nails the learner to a user interface of a machine, laptop, tablet or smart phone. The idea is to transform our surroundings as the learner interface to knowledge, and conceptually adopt the Ambient Intelligence notion to education. Ambient Intelligence (AmI) is the potential of injecting AI into our immediate environment be it a room, an office, a house, a boat, car or plane. The ambient around us will intelligently figure out, sensitise and duly respond to our specific needs and requirements in an effort to improve and add value to the way we experience the world around us. Apart from the pervasiveness nature of AmI as well as its ubiquitous advantages, it brings additional qualities like embeddedness, context awareness, personalization, adaptivity, and anticipatory behavior. Some of these qualities form part of the proposed model while all the rest can be applied to a learning environment in which case will be the immediate environment around the learner. To further link the e-learning model presented to this AmI learning environment (AmILE) the learner needs to be connected online as well as willing to be actively involved within an educational experience. Important to point out that as the academic requirements are being set and optimally defined to suit the educational needs and objectives of this AmILE the technological requirements start to materialise to accommodate AmILE rather than the other way round. Additional technologies that will be highlighted in the next section will also have an integral role within such a scenario and which conveniently come in handy to achieve a common academic goal as they compatibly come together. This leads to another parallel drawn from the proposed iPLE model whereby a number of techniques are merged to address a number of concerns within an epistemological framework. The AmILE would entail highly intelligent techniques to coordinate the functionality of numerous physical and implicit sensors together with capabilities of software-triggering and physical actuators. This environment frees the iPLE from the browser and enables the ambient around the learner taking e-learning to new heights and multi-sensory dimensions. An initial study kicks-off on October 2017 and will be subject of numerous empirical studies similar to the 2016 iPLE ones that have provided material for this book.

7.3 Future Trends

Additional technologies that have also been notoriously identified by numerous researchers [2–6] and that have previously been earmarked as potential future key drivers like AI, smart mobile technologies, distributed computing, and intelligent filtering start to fall in place within the ambient intelligent learning environment, as well as others like Internet of Things (IoT) and wireless communications will also feature as essential technologies to bring together these numerous technologies. Additional future trends within the e-learning evolution can easily be integrated within the AmILE scenario. To mention a few:

- 3-D visualizations can easily be projected within any environment the learner is in. Realistic imagery further enhance the level of immersion and the effectiveness of the e-learning environment. Various modes of interaction are possible that could require additional hardware in cases like Virtual Reality and Augmented Reality, while others like Mixed Reality would need a specially equipped environment which could easily be a dedicated educational space as part of the AmILE scenario;
- Pedagogical agents are purely AI components that employ personalization techniques to customise the learning process. The iPLE is a single pedagogical agent in itself that works in isolation to assist a particular learner, but collaborates with other agents through the PLN. Multi pedagogical agents would ideally cater for the multi-dimensionality of the AmILE scenario where the complexity of integrating different PLNs from around the learners ambient will be required;
- Biometric recognition is an ideal technology to unobtrusively identify the uniqueness of one person and distinguish this same person from all others. Such a technique would implicitly recognise a learner without any human intervention and assist the PLP component to gather, process and generate an accurate learner profile;
- Gamification is another e-learning or methodology that is already setting in as an effective and engaging activity in its own right. The interactive and challenging elements that enable learners turned gamers to master different levels, immerse themselves, and develop strategies are ideal learning settings within any ambient. The AmILE needs to be sensitive to what kind of games and which genre is most effective with each specific learner, if any. On the other hand the intelligent ambient is required to lend itself adequately or provide a dedicated environment to augment the effectiveness of this methodology.

Additional future trends might emerge in the next few years which would potentially lend themselves to the AmiLE scenario, however the conceptual objectives have been set and are in no way open to accommodate the novel technology or trend. On the contrary if the AmILE project triggers off and actuates the need of some novel technique or trend then the technology will be employed and applied to accommodate the educational needs of the next e-learning generation.

7.4 Conclusion

AI-injected applications perform tasks that usually require a human to do and fundamentally simulate human intelligence to the extent that it is not possible to distinguish whether communication is enacted by a human or an intelligent application. Teaching is a special task that no AI application, intelligent as it may be, can replace it if performed optimally. The role of a talented, inspirational, resourceful and skillful educator cannot be replaced or replicated by any other practice or approach however this is not always the case. Blended learning manages to strike a healthy balance between the electronic and the human contributions however this is not always possible and numerous e-learning systems rely entirely on electronic communication to address the needs of numerous learners. In such cases the electronic medium needs to be optimised and efficiently delivered to be as effective as possible to successfully deliver an education as close as possible to an ideal face-to-face delivery. Over the years e-learning systems and online academic platforms strived and endeavored in closing the gap between the cold electronic medium and the tender physical contact. Researchers and educational technologists as well as e-learning advocates sought to improve and continuously forge ahead evolving from one generation to the other as technologies became available. This book presents an attempt to set the pace towards the next generation of e-learning systems within a proposed model that feature the combination of AI techniques, Web 2.0 functionalities, and improved personalization capabilities. The model is grounded in sound and compatible learning theories and addresses a number of e-learning issues that have scourged e-learning models over the years. The actual development and deployment of the iPLE [7] together with the empirical study in July 2015 to test and fully evaluate the proposed platform was not within the scope of the book. The conceptual philosophy behind the model, coupled with personal epistemological positioning enabled the successful composition of an intelligent personal learning environment that essentially personalises the e-learning services that are currently available. This book does not only recapitulate all the hard work performed these last five years, and nor does it characterise the end of an exhilarating journey, but merely demarcates the beginning of a promising way forward as new research avenues are uncovered characterizing the future of online education and intelligent e-learning platforms while promising to improve and enhance peoples interaction and attitude towards e-learning and online education in general.

References

1. Rubens, N., Kaplan, D., Okamoto, T.: E-Learning 3.0: anyone, anywhere, anytime, and AI in dickson C.K.W. In: Minhong, W., Elvira, P., Qing, L., Rynson, L. (eds.) New Horizons in Web Based Learning: ICWL 2011 International Workshops, KMEL, ELSM, and SPeL, Hong Kong, pp. 171–180. Springer, Berlin (2014)

2. Miranda,P., Isaias, P., Costa, C.J.: E-Learning and Web Generations: Towards Web 3.0 and E-Learning 3.0, 4th International Conference on Education, Research and Innovation, vol. 81. IACSIT Press, Singapore. (2014)
3. Hussain, F.: E-Learning 3.0 = E-Learning 2.0 + Web 3.0? IADIS International Conference on Cognition and Exploratory Learning in Digital Age (CELDA 2012) pp. 11–18. (2012)
4. Amarin, N.Z.: Web 3.0 and its reflections on the future of e-learning. Acad. J. Sci. 115–121 (2015)
5. Wheeler, S.: e-Learning 3.0 (2009). http://www.steve-wheeler.co.uk/2009/04/learning-30.html Cited 12 March 2017
6. Hussein, M.: Transition to Web 3.0: E-Learning 3.0 opportunities and challenges. EELU ICEL 2014 International conference on e-learning. Cairo, Egypt. (2014)
7. Montebello, M.: Personalised e-Learning, Ed.D. Thesis, University of Sheffield, UK. (2016)

Glossary

Adaptive Learning Theory A learning theory that conceptualises the use of technology to customise and tailor educational resources to accommodate the specific and unique needs of each learner.

Ambient Intelligence - AmI Artificial intelligence applied to an enclosed environment like a room, an office, a house, a boat, car, plane, or even a city. The ambient that surrounds the user or learner intelligently recognises the uniqueness of the person and reacts accordingly in a personalised manner.

Artificial Intelligence - AI The use of computer science techniques to develop computer programs in an attempt to simulate human behaviour. These programs perform tasks that usually require a human to do and thereby convey a sense of added value when compared to simple computer tasks.

Connectivism Learning Theory A theory first put forward by Siemens [1] presupposes that in the digital information age knowledge is the product of influences from a number of sources, both human and non-human. When an individual is able to reconcile all the connections from the various information sources in a meaning-making exercise, learning happens.

Crowdsourcing The use of online users to collectively contribute and aggregate information towards a common goal. Initially coined by Jeff Howe and Mark Robinson to describe the way commercial entities outsourced tasks to the crowd over the World-Wide Web [2].

e-Learning Is learning on Internet Time, the convergence of learning and networks. e-Learning is a vision of what corporate training can become. E-Learning is to traditional training as eBusiness is to business as usual [3]. Different versions and generation of e-learning exist as technologies evolved over the years.

Internet of Things - IoT An electronic network of all objects in the real world connected and uniquely identified with the ability to communicate and coordinate amongst themselves. The smart devices could also be embedded in everyday objects that enables them to transmit and receive data.

© Springer International Publishing AG 2018
M. Montebello, *AI Injected e-Learning*, Studies in Computational Intelligence 745,
https://doi.org/10.1007/978-3-319-67928-0

Learning Technologies Different media, technology-based applications and tools that can be used to facilitate and support learning. Learning technologies also include the 21st century digital practices that would require a specific set of skills and attitudes.

Pedagogy The art and science of teaching. Usually taken for granted however effective teaching requires specific skills and experience. Educators can employ a plethora of teaching strategies to optimise the use of the learning medium selected.

Personal Learning Environment - PLE A combination of personal academic tools, services and communities that a learner makes use of. Electronic personal learning spaces are traditionally made up of two components, namely, a personal learning network and a personal learning portfolio.

Personal Learning Network- PLN A virtual and informal network of friends and resources that a learner can interact with and from which information and knowledge is extracted for personal use. A personal learning network usually forms part of a personal learning environment.

Personal Learning Portfolio - PLP A compendium of academic works that act as educational evidence of a particular learner. It is commonly part of a personal learning environment and is used to assess the learner, keep an academic record, and act as feedback to the learner.

Self-Determinism Learning Theory A learning theory that promotes the motivation of the self within a learning environment. Deci and Ryan [4] initial theory about intrinsic and extrinsic motivation and basic psychological needs applied to the educational domain.

Social Constructivism Learning Theory A theory posited by Vygotsky [5] that describes how meaning making can be aided by the social context in which the learner is found. Therefore, community and collaborative activities become an important influence on the learning.

Social Networks This term refers to the connections between individuals in a community. Christakis and Fowler [6] define this as an organised set of people that consists of two kinds of elements: human beings and the connections between them- Real, everyday social networks evolve organically from the natural tendency of each person to seek out and make many or few friends, to have large or small families, to work in personable or anonymous workplaces (p. 13).

Technology Acceptance Model - TAM Based on the Davis [7] theory of reasoned action it models how learners come to accept, usefulness and ease of use, a system like an e-learning environment.

Virtual Learning Environment - VLE This term broadly encompasses virtual spaces that are used for learning. Such environments can include Learning Management Systems (LMS), Multiuser Virtual Environments (MUVEs), Virtual Worlds (VWs), and Serious Games.

Web 2.0 OReilly [8] coined this term to demarcate a phase within the evolution of the WWW whereby websites allow user-generated content thus encouraging web user to author, contribute, share, and distribute their own and others material. Social media were a direct result of this particular phase that also has dynamic characteristics in contrast to previous static read-only counterparts.

World-Wide Web - WWW The massive knowledge base of information spread over the global network of servers known as the Internet. Different generations of WWW represent the evolution of how this technology has radically changed over a short period of time from a read-only, to a read-write and share.

References

1. Siemens, G.: Connectivism: A Learning Theory for the Digital Age (2004). http://www. elearnspace.org/Articles/connectivism.htm Cited 12 Feb 2013
2. Howe, J., Robinson, M.: The Rise of Crowdsourcing in Wired (2006). https://www.wired.com/ 2006/06/crowds/ Cited 23 Feb 2015
3. Cross, J.: An informal history of eLearning. On Horiz. **12**(3), 103–110 (2004). Emerald
4. Deci, E., Ryan, R.: Intrinsic Motivation and Self-Determination in Human Behavior. Plenum Press, New York (1985)
5. Vygotsky, L.: Interaction between learning and development. In: Gauvain, M., Cole, M. (eds.) Readings on the Development of Children, pp. 29–36. W.H. Freeman and Company, New York (1997)
6. Christakis, N., Fowler, J.: Connected: The Amazing Power of Social Networks and How They Shape Our Lives. HarperCollins Publishers, London (2011)
7. Davis, F.: User acceptance of information technology: system characteristics, user perceptions and behavioral impacts. Int. J. Man-Mach. Stud. **38**(3), 475–487 (1993)
8. O'Reilly, T.: What Is Web 2.0 Design Patterns and Business Models for the Next Generation of Software (2005). http://www.oreilly.com/pub/a/web2/archive/what-is-web-20.html?page=1 Cited 18 Jan 2015

Author Index

© Springer International Publishing AG 2018
M. Montebello, *AI Injected e-Learning*, Studies in Computational Intelligence 745,
https://doi.org/10.1007/978-3-319-67928-0

Subject Index

© Springer International Publishing AG 2018
M. Montebello, *AI Injected e-Learning*, Studies in Computational Intelligence 745,
https://doi.org/10.1007/978-3-319-67928-0

Printed in the United States
By Bookmasters